THE
A–Z OF
ERROR-FREE
RESEARCH

THE
A–Z OF
ERROR-FREE
RESEARCH

Phillip I. Good

CRC Press
Taylor & Francis Group
Boca Raton   London   New York

CRC Press is an imprint of the
Taylor & Francis Group, an **informa** business

A CHAPMAN & HALL BOOK

Cover credit: Gary C. Carlsen, D.D.S.

CRC Press
Taylor & Francis Group
6000 Broken Sound Parkway NW, Suite 300
Boca Raton, FL 33487-2742

© 2013 by Taylor & Francis Group, LLC
CRC Press is an imprint of Taylor & Francis Group, an Informa business

No claim to original U.S. Government works

Printed in the United States of America on acid-free paper
Version Date: 20120725

International Standard Book Number: 978-1-4398-9737-9 (Paperback)

Library of Congress Cataloging-in-Publication Data

Good, Phillip I., author.
  The A-Z of error-free research / Phillip I. Good.
  pages cm
  Includes bibliographical references and index.
  ISBN 978-1-4398-9737-9 (pbk.)
  1. Research--Statistical methods. 2. Science--Methodology. I. Title.

Q180.55.S7G66 2013
001.4'2--dc23
2012021067

Visit the Taylor & Francis Web site at
http://www.taylorandfrancis.com

and the CRC Press Web site at
http://www.crcpress.com

Contents

PART II Data Collection

PART III Analyzing Your Data

PART V   Reporting Your Results

PART VI Nonrandom Samples

Preface

This new text is designed to assist you in the transition from slavish student to bold independent researcher. Use it as a self-help refresher course, as a textbook for a course in research methods, and as a second course in statistics. It contains step-by-step prescriptions, numerous worked-through examples, and the R code you'll need to implement the methods to aid in making use of the material.

You will learn when to use statistics, the best ways to cope with variation, how to design an experiment, how to determine optimal sample size, and how to collect usable data for experiments, clinical trials, and surveys. You will be guided to the best statistical procedures for your current application and taken step by step through model development and the reporting of your results for publication.

Chapter 1 provides an overall prescription along with guidance as to when, and when not, to use statistics.

Chapters 2, 3, and 4 walk us through the planning phase.

Chapters 5 and 6 take us step by step through data collection. Methods for sample size determination are deferred to Chapter 11.

A comprehensive guide to contemporary methods in data analysis is provided in Chapter 7 through Chapter 10.

Techniques for developing models that will provide a basis for future research are given in Chapters 12, 13, and 14.

Chapters 15 through 17 describe reporting techniques that will ensure your research efforts get the credit they deserve.

Chapter 18 is devoted to case-control and cohort studies.

Put this book to practical use today. Your comments are welcomed.

Phillip I. Good
drgood@statcourse.com

Acknowledgments

My thanks to Tony Rowe for his comments on a preliminary version of this manuscript.

About the Author

Phillip I. Good is a Canadian-American mathematical statistician. He was educated at McGill University and the University of California at Berkeley.

He was among the first to apply the bootstrap in his 1975 analyses of 2 × 2 designs with a missing cell.[*] His chief contributions to statistics are in the area of small sample statistics, including a uniformly most powerful unbiased (UMPU) permutation test for Type I censored data,[†] an exact test for comparing variances, and an exact test for cross-over designs.[‡]

His published texts in statistics include the following:

- *A Practitioner's Guide to Resampling for Data Analysis, Data Mining, and Modeling* (Chapman & Hall/CRC, 2011)
- *Analyzing the Large Number of Variables in Biomedical and Satellite Imagery* (Wiley, 2011)
- *Managers' Guide to the Design and Conduct of Clinical Trials* (Wiley, 2002; 2nd ed., 2006)
- *Introduction to Statistics Using Resampling Methods and R/S-Plus* (Wiley, 2005)
- *Introduction to Statistics Using Resampling Methods and Excel* (Wiley, 2005)
- *Common Errors in Statistics (and How to Avoid Them)* (with J. Hardin, Wiley, 2003; 4th ed., 2012)
- *Applying Statistics in the Courtroom: A New Approach for Attorneys and Expert Witnesses* (Chapman & Hall, 2001)

---

[*] Makinodan, T., J.W. Albright, C.P. Peter, P.I. Good, and M.L. Heidrick. Reduced humeral immune activity in long-lived mice. *Immunology.* 1976; 31(76): 400.

[†] Good, P.I. Globally almost most powerful tests for censored data. *Journal of Nonparametric Statistics.* 1991; 1(92): 253–262.

[‡] Good, P. and F. Xie. Analysis of a crossover clinical trial by permutation methods. *Contemporary Clinical Trials.* 2008; 29: 565–568.

- *Resampling Methods* (Birkhauser, 1999; 3rd ed., 2005)
- *Permutation, Parametric, and Bootstrap Tests of Hypotheses* (Springer-Verlag, 1994; 3rd ed., 2005)

He has published articles in

- Biology: *Exp. Geron.* 8(73): 147; *Mech. Age. Devel.* 4(75): 339.
- Biomathematics: *J. Theoret. Biol.* 34(72): 99; *Bull. Math. Biol.* 38(76): 295.
- Biostatistics: *Contemporary Clinical Trials* 29(08): 565.
- Computer Science: *Computer Architecture News* 16(88)3: 40.
- Physics: *Physics Essays* 23(10): 368.
- Probability: *J. Aust. Math. Soc.* 8(68): 716.
- Statistics: *J. Nonpar. Statist.* 1(92): 253–262; *J. Modern Appl. Statist. Meth.* 1(02): 243.

Good has published nine texts on the application of microcomputers in business, as well as several hundred articles on microcomputers in over a dozen different computer magazines. He is the writer of some 21 novels, two novellas, and two collections of short stories published by zanybooks.com.

CHAPTER 1

Research Essentials

We cannot help remarking that it is very surprising that research in an area that depends so heavily on statistical methods has not been carried out in close collaboration with professional statisticians.

From the report of an independent panel looking into "Climategate"

Belief is essential to the religious experience. The law is what judges say it is. But knowledge arises only from observation and deductive mathematics.

Norman Friedgut

In this chapter, as throughout this text, we provide first a preventive prescription. If this prescription and the more detailed prescriptions in subsequent chapters are followed carefully, you will be guided to the correct, proper, and effective use of statistics in your research while avoiding the pitfalls.

Prescription

Success in your research efforts will follow only if five essentials are observed:

1. Have a clear understanding of your objectives before you begin. Set forth your objectives and your research intentions *before* you conduct a laboratory experiment, a clinical trial, a survey, or analyze an existing set of data.
2. Know what you want to measure, and then monitor your efforts to ensure the data you are collecting is the data you planned to collect.
3. Describe and analyze your data appropriately and effectively.
4. Develop models from your data that can be used as the basis of subsequent data collection and analysis.
5. Don't be afraid of success. Report your results in so comprehensive a fashion that they can be readily replicated and built upon by others.

Oops, there is a sixth essential for the journeyman researcher:

6. Never collect data in isolation but as part of an ongoing program of research. Ideally, the results of each of your research efforts will serve as a starting point for your next one.

Fundamental Concepts

Three concepts are fundamental to the design of clinical trials, experiments, and surveys: variation, population, and sample.

A thorough understanding of these concepts will forestall many errors in the collection and interpretation of data.

If there were no variation, if every observation were predictable, a mere repetition of what has gone before, there would be no need for statistics.

Variation

Variation is inherent in virtually all our observations. We would not expect outcomes of two consecutive spins of a roulette wheel to be identical. One spin may result in red; a second, in black. The outcome varies from spin to spin.

Certain gamblers watch and record the spins of a single roulette wheel hour after hour hoping to discern a pattern. A roulette wheel is a mechanical device, so perhaps a pattern will emerge. But even those observers do not anticipate finding a pattern that is 100% predetermined. The outcomes are just too variable depend on too many factors, only a few of which are directly observable.

Anyone who spends time in a schoolroom, as a parent or as a child, can see the vast differences among individuals. This one is tall and that one short, though all are the same age. My headache is gone with half an aspirin, but my wife requires four times that dosage.

There is variability even among observations satisfying deterministic formulas, such as the position of a planet in space or the volume of gas at a given temperature and pressure. Position and volume satisfy Kepler's laws and Boyle's law, respectively (the latter over a limited range of values), but the observations we collect will depend upon the measuring instrument (which may be affected by the surrounding environment) and the observer. Cut a length of string and measure it three times. Do you record the same length each time?

In designing an experiment or survey, we must always consider the possibility of errors arising from the measuring instrument and from the observer. It is one of the wonders of science that Kepler was able to formulate his laws at all given the relatively crude instruments at his disposal.

Deterministic, Stochastic, and Chaotic Phenomena

A phenomenon is said to be deterministic if given sufficient information regarding its origins, we can successfully make predictions regarding its future behavior.

But we don't always have all the necessary information. Planetary motion falls into the deterministic category once one makes adjustments for *all* gravitational influences, the other planets as well as the sun.

Nineteenth-century physicists held steadfast to the belief that all atomic phenomena could be explained in deterministic fashion. Slowly, it became evident that on the subatomic level many phenomena were inherently stochastic in nature; that is, while one could specify a probability distribution of possible outcomes, no specific outcome was certain.

Strangely, 21st-century astrophysicists continue to reason in terms of deterministic models. They add parameter after parameter to the lambda cold dark matter model, hoping to improve its goodness of fit to astronomical observations. Yet, if the universe we observe is only one of many possible realizations of a stochastic process, goodness of fit offers absolutely no guarantees of the model's applicability (see, for example, Good, 2010).

Chaotic phenomena differ from the strictly deterministic in that they are strongly dependent on initial conditions. A random perturbation from an unexpected source (the proverbial butterfly's wing) can result in an unexpected outcome. The growth of cell populations has been described in both deterministic (differential equations) and stochastic terms (birth and death process), but a chaotic model (difference lag equations) is more accurate.

Population

From time to time, someone will ask us how to generate confidence intervals for the statistics arising from a total census of a population. Our answer is no, we cannot help. Population statistics (mean, median, 30th percentile, dispersion) are not estimates. They are fixed values and will be known with 100% accuracy if two criteria are fulfilled:

1. Every member of the population is observed.
2. All the observations are recorded correctly.

Kepler's "laws" of planetary movement are not testable by statistical means when applied to the original planets (Jupiter, Mars, Mercury, and Venus) for which they were formulated. But when we make statements such as, "Planets that revolve around Alpha Centauri will also follow Kepler's laws," then we begin to view our original *population*, the planets of our sun, as a *sample* of all possible planets in all possible solar systems.

A major problem with many studies is that the population of interest is not adequately defined before the sample is drawn. Don't make this mistake. A second major problem is that the sample proves to have been drawn from a different population than was originally envisioned. We discuss possible biases in the chapters on experimental design and data collection

Samples

A fundamental principle of statistics is that a great deal may be learned about a population by taking a representative sample from it. A second, equally fundamental principle is that as the sample grows larger, it will more closely resemble the population from which it is derived.

Early in the 20th century, statisticians in industry and government began to make use of samples to describe and estimate the characteristics of the populations from which the samples were drawn. At almost the same time, the manufacturers of poorly designed and unsafe products began to protest against the use of statistics.

In 1946, the U.S. Food and Drug Administration (FDA) condemned a shipment of prophylactics as adulterated and misbranded after tests in which 6 of 82 of one brand examined in post-seizure tests, and 8 of 108 of another examined pre-seizure contained holes. The manufacturer protested that it was unfair to condemn the entire shipment; they also protested the tests themselves because they were destructive in nature. The court noted that government regulations referred specifically to samples and the need to make them available for testing and ruled in the FDA's favor.*

But not every sample will do. Two criteria must be satisfied:

1. The sample must be selected at random from the population.
2. The sample must be representative of the population.

Small samples may give a distorted view of the population. For example, if a minority group comprises 10% or less of a population, a jury of 12 persons selected at random from that population will fail to contain any members of that minority at least 28% of the time.

As a sample grows larger, or as we combine more clusters within a single sample, the sample will grow to more closely resemble the population from which it is drawn. How large a sample must be to obtain a sufficient degree of closeness will depend on the manner in which the sample is chosen from the population.

Are the elements of the sample drawn at random, so that each unit in the population has an equal probability of being selected? Are the elements of the sample drawn independently of one another? If either of these criteria is not satisfied, then even a very large sample may bear little or no relation to the population from which it was drawn.

An obvious example is the use of recruits from a Marine boot camp as representatives of the population as a whole or even as representatives of all Marines. In fact, any group or cluster of individuals who live, work, study, or pray together may fail to be representative for any or all of the following reasons (Cummings and Koepsell, 2002):

---

* United State v. 43 1/2 gross rubber prophylactics, 65 F. Supp 534 (MN 4th div 1946).

1. Shared exposure to the same physical or social environment;
2. Self-selection in belonging to the group;
3. Sharing of behaviors, ideas, or diseases among members of the group.

A sample consisting of the first few animals to be removed from a cage will not satisfy these criteria either, because depending on how we grab, we are more likely to select either more active or more passive animals. Activity tends to be associated with higher levels of corticosteroids, and corticosteroids affect virtually every body function.

Precautions

To forestall sample bias in your studies, before you begin, determine all the factors that can affect the study outcome (gender and lifestyle, for example). Subdivide the population into strata (males, females, city dwellers, farmers), and then draw separate samples from each stratum. Ideally, you would assign a random number to each member of the stratum and let a computer's random number generator determine which members are to be included in the sample.

Being selected at random does not mean that an individual will be willing to participate in a public opinion poll or some other survey. But if survey results are to be representative of the population at large, then pollsters must find some way to interview nonresponders as well. This difficulty is exacerbated in long-term studies, as subjects fail to return for follow-up appointments and/or move without leaving a forwarding address. Again, if the sample results are to be representative, some way must be found to report on subsamples of the nonresponders and the dropouts.

"Statistics *is* required if and only if variation is inherent in the data. If every observation is predictable, a mere repetition of what has gone before, there is no need for statistics."

Will the Data Require Statistical Methods?

It's all too easy to load statistics software into our computer today and begin to analyze our data. Yet a moment or so of reflection often can replace several thousand calculations.

During the Second World War, a group was studying planes returning from bombing Germany. They drew a rough diagram showing where the bullet holes were and recommended those areas be reinforced. A statistician, Abraham Wald (1980),[*] pointed out that essential data were missing from the sample they were studying: What about the planes that didn't return from Germany?

When we think along these lines, we see that the two areas of the plane that had almost no bullet holes (where the wings and where the tail joined the fuselage) are crucial. Bullet holes in a plane are likely to be at random, occurring over the entire plane. Their absence in those two areas in returning bombers was diagnostic. No statistics are required.[†]

Too often, social scientists will attempt to detect relationships via a factor analysis, structural equations models, or a multiple regression. As spurious correlations are all too common,[‡] wouldn't their time be better spent thinking though the relationships?

You may well be tempted to report on unexpected patterns you observe in your data. Feel free to do so, *provided* you do not attempt to ascribe a degree of certainty to what you have observed.

Patterns in data can suggest, but cannot confirm hypotheses unless these hypotheses were formulated *before* the data were collected.

Everywhere we look, there are patterns. In fact, the harder we look, the more patterns we see. Three rock stars die in a given year. Fold the United States twenty-dollar

---

[*] This reference may be hard to obtain. Alternatively, see Mangel and Samaniego, 1984.
[†] Though, in this example as in many others, the input of a statistician was essential.
[‡] See Freedman, 1983.

bill in just the right way and not only the Pentagon but the Twin Towers in flames are revealed. It is natural for us to want to attribute some underlying cause to these patterns. But the laws of probability tell us that more often than not patterns are simply the result of random events.

How can we determine whether an observed association represents an underlying cause-and-effect relationship or is merely the result of chance? We need conduct a second set of trials in accordance with the methods discussed in the next few chapters.

Summary

Careful attention to the essentials described in this chapter will result in making your research efforts more productive, and, hopefully, with greater impact on humanity's understanding of the laws of nature.

The population(s) of interest must be clearly defined before we begin to gather data. Our samples must be both random and representative. We must never compute p-values for hypotheses derived after viewing the data.

Statistical procedures for hypothesis testing, estimation, and model building are only a *part* of the decision-making process. They should never be quoted as the sole basis for making a decision (yes, not even those procedures that are based on a solid deductive mathematical foundation). As philosophers have known from the time of Hume and Berkeley, extrapolation from a sample or samples to a larger incompletely examined population always must entail a leap of faith.

PART I

PLANNING

Prescription

Before you complete a single data collection form

1. Set forth your objectives and the use you plan to make of your research (see Chapter 2).
2. Gather qualitative data (see Chapter 2).
3. Formulate your hypothesis and all of the associated alternatives. Define your end points. List possible experimental findings along with the conclusions you would draw and the actions you would take for each of the possible results (see Chapter 2).
4. Specify the outcomes of interest and the possible surrogates (see Chapters 2 and 3).
5. List all possible sources of variation (see Chapter 3).
6. Decide how you will cope with each source (see Chapter 3).
7. Define the populations to which you will apply the results of your analysis (see Chapters 3 and 4).

8. Describe in detail how you intend to draw a representative random sample from the population. (see Chapter 4). Because case-control and cohort studies differ in so many respects—from how samples ought to be collected, to how they ought to be analyzed, to how their results ought to be reported—they are considered in Chapter 18.

9. Describe how you will ensure the independence of your observations (see Chapter 4).

10. Describe how you will assign treatments to the members of the sample (see Chapter 4).

CHAPTER **2**

Hypotheses and Losses

Prescription

- State Your Objectives
- Gather Qualitative Data
- Formulate Your Hypotheses
- List Consequences of Possible Decisions

State the Objectives of Your Research

An explicit written statement of your objectives generally will lead to their narrowing. Once one writes "Cure Cancer" on paper, its quixotic nature is immediately apparent. "Cure prostate cancer," or better, "Halt the growth of a prostate cancer" seem more achievable. Given the scrutiny that experiments with humans are subjected to, you might then start searching for a good animal model or perhaps even consider the possibility of growing prostate tumor cells in vitro.

A general statement of objectives that may be used as a template for your own studies might take the following form: "The purpose of this study is to demonstrate that

- in treating conditions A, B, C
- with subjects having characteristics D, E, F

- an intervention of the form G
- is equivalent to/as effective as/as or more effective than an intervention of the form H
- and has fewer unwanted effects.

When completing such a template, you will need to make explicit what is meant by "effective" and to list some, if not all, of the unwanted effects you hope to diminish or eliminate.

For Surveys

For surveys, begin with a brainstorming session and create a list of topics that you want to know more about. As you create the list, don't worry about phrasing or overlap of topics; you'll be taking multiple passes through this list. Gradually refine and build upon the initial list. Look for items that are vague or abstract and consider how they might be made more concrete. Prioritize.

Gather Qualitative Data

Suppose you wish to investigate why couples who have come to the brink of divorce because of a severe transgression on the part of one of them are unable to forgive each other and repair the relationship. This area of research being relatively new (to you), you are not yet able to formulate a hypothesis. Instead, you do all of the following:

- Consult your textbooks
- Google (or Yahoo!) the topic
- Browse the library
- Conduct interviews with a series of couples using questions that are as open-ended as possible in an attempt to keep your opinions/biases from influencing the direction of the couples' answers.

Lise DeShea notes that "Qualitative studies can inform quantitative pursuits. For example, people struggling with the forgiveness process sometimes make a decision to forgive, yet emotionally they really haven't. Ev Worthington, one of the most prolific forgiveness researchers, observed this dichotomy and went on to create two scales for quantitative research, one measuring Decisional Forgiveness and the other measuring Emotional Forgiveness."

Jay Warner's objective was to keep poorly built joints from being shipped to customers. He describes collecting nominal and ordinal information about an aluminum braze joint. A visual assessment of size and uniformity was assigned to an ordinal scale of 1–5, which then was correlated to performance in a customer application. Color turned out to be a poor indicator, and was dropped from further consideration as a predictor.

My own view is that of an Earth-born anthropologist who has just landed on Rocannon's World: It's best to look around carefully before jumping to hypotheses.

Formulating Hypotheses

A well-formulated hypothesis will be both quantifiable and testable, that is, involve measurable quantities or refer to items that may be assigned to *mutually exclusive* categories.

A well-formulated statistical hypothesis takes one of the forms "Some measurable characteristic of a population takes one of a specific set of values" or "Some measurable characteristic takes different values in different populations, the difference(s) taking a specific pattern or a specific set of values."

While we can begin by stating our hypotheses in qualitative form as with "A meteor hitting the Earth resulted in a drop in temperature that killed the dinosaurs," and "A glass of Gatorade between heats will increase a runner's speed." We need to finish by stating our hypotheses explicitly in quantitative terms: "A 4-ounce glass of Gatorade between heats will decrease the time a male runner requires to run a half mile by an average of at least 4.5 seconds."

Note my use of the word *average*. We can never specify a single-fixed outcome for a variable but only some
aspect of the variable's distribution F, such as its arithmetic mean (often referred to less precisely as its average) or expectation, its median (or 50th percentile), or its
standard deviation.

In case you have forgotten, the cumulative distribution function $F_Y[x]$ of a variable Y is the probability that
the variable Y takes a value less than or equal to x. The
cumulative distribution function (cdf) of the heights of
sixth-grade girls is depicted in Figure 2.1. If a variable Y
can take a total of K possible values x_1, x_2, through x_K with
probabilities p_1, p_2, though p_K, respectively, its expectation
or arithmetic average A_Y is the sum $x_1{}^*p_1 + x_2{}^*p_2 + \ldots +
x_k{}^*p_K$, written $\sum_{k=1}^{K} x_k * p_k / K$. Its median M or 50th
percentile is such that $F_Y[M] = 50\%$. And its standard

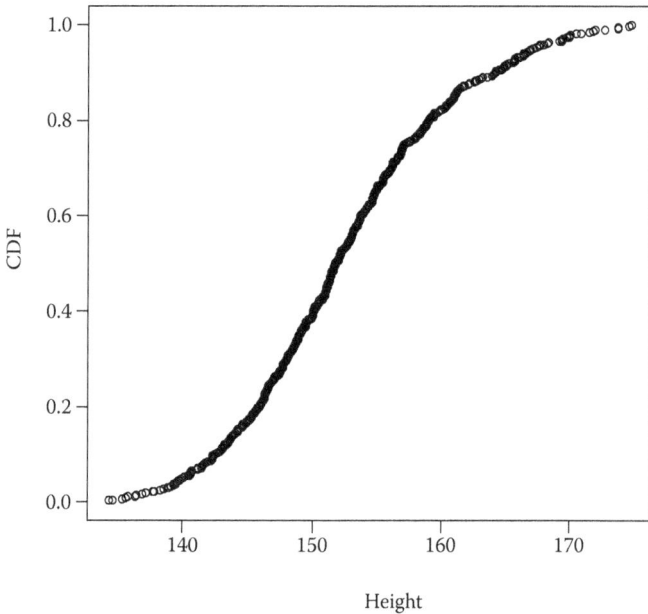

FIGURE 2.1 Cumulative distribution function of the heights of sixth-grade
girls in centimeters.

deviation is the square root of the sum of its deviations about its mean, written $\sum_{k=1}^{K} \left(x_k - A_Y \right)^2 * p_k / K$.

A testable hypothesis must be specified in terms of some aspect of a variable's distribution. All of the following are testable hypotheses:

> The mean height of sixth-grade girls in a California classroom is 55 centimeters.
> The median height of sixth-grade girls in a California classroom is 54 centimeters.
> The heights of sixth-grade California girls are normally distributed.*

Warning: In order to test a hypothesis by statistical means, it must be specified before the data are examined. Failing to do so is equivalent to the magician's trick of "forcing a card." If you do see something interesting in a data set, use this observation as the basis of a future study.

The hypothesis must be numeric in form and must concern the value of some population parameter. Examples: More than 50% of those registered to vote in the state of California prefer my candidate. The arithmetic average of errors in tax owed that are made by U.S. taxpayers reporting $30,000 to $50,000 dollars income is less than $50. The addition of Vitamin E to standard cell growth medium will increase the lifespan of human diploid fibroblasts by at least a generation.

Oops, the expected effect must be of practical interest, else why conduct the study. "The addition of Vitamin E to standard cell growth medium will increase the lifespan of human diploid fibroblasts by more than 29 generations."

In these examples, we've tried to specify the population from which the samples will be taken as precisely as possible.

---

* As a percentage of girls will experience a growth spurt during the sixth grade, their heights are a mixture of two normal distributions.

Specify the Decisions and Associated Costs

What decisions are associated with the outcome of your study? If you've tested the addition of Vitamin E to standard cell growth medium and found that it will increase the lifespan of human diploid fibroblasts by no more than 29 generations, will you abandon this line of research? Or will you repeat your tests with a second larger sample? If your survey shows that 9% of your county's population is below the poverty line, will you breathe a sigh of relief? Repeat the survey with a larger sample? Redefine poverty in accord with recent increases in the standard of living?

Before you conduct a study, you need to do all of the following:

- List all potential outcomes. (Defining several nonoverlapping ranges of values will suffice. In our Vitamin E experiment, these might include "under 20 generations," "20–29," "30–40," and "over 40.")
- List the decisions you will make for each possible outcome (see Table 2.1).
- Prepare a cost matrix associated with these decisions. This matrix will help you in determining sample size as well as in deciding whether it is worth doing the study in the first place (see Table 2.2).

TABLE 2.1   Decision Matrix

% Unwanted Side Effects Observed	>1%	1%	0
Abandon Tests and Drug	X		
Repeat Tests		X	
Bring Drug to Market			X

TABLE 2.2   Cost Matrix

	Drug Safe	Drug Unsafe
Abandon Drug	Missed opportunity Public denied access to effective treatment	
Repeat Tests	Time-consuming Delays access to effective treatment	Time-consuming
Bring Drug to Market		Consumers suffer/die. Families suffer/die. Researcher sued.

Specify the Alternatives

Almost all contemporary methods of hypothesis testing, their rejection regions, cut-off values, and resultant p-values are based on the likelihood ratio

$$\frac{\text{Probability of the observed outcomes under the alternative hypothesis}}{\text{Probability of the observed outcomes under the primary hypothesis}}$$

To derive a test, we order each of the possible outcomes in accordance with this ratio. When the likelihood ratio is large, we say the outcome rules in favor of the alternative hypothesis. Working downwards from the outcomes with the highest values, we continue to add outcomes to the *rejection* region of the test—so called because these are the outcomes for which we would reject the primary hypothesis—until the total probability of the rejection region under the primary hypothesis is equal to some predesignated *significance level*.

TABLE 2.3 Probabilities of Outcomes under the Primary Hypothesis and an Alternative

	Really hate	Dislike	Indifferent	Like	Really like
Primary Hypothesis	10%	25%	45%	10%	10%
Alternate Hypothesis	5%	10%	30%	30%	25%
Likelihood Ratio	1/2	1/2	3/4	3/2	5/2

Based on the probabilities in Table 2.3, we would reject the primary hypothesis at the 10% level only if the test subject really liked a product

To see that we have maximized the power of our test at a given significance level by using the likelihood ratio as our criteria, suppose we replace one of the responses we assigned to the rejection region with one we did not. From the row of Table 2.3 corresponding to the primary hypothesis, we see that if we are not to exceed the specified significance level, the probability the newly selected response would occur under the null hypothesis must be less than or equal to the probability of the outcome it replaced. We could reject if the respondents "liked" the product.

Because of how we assigned outcomes to the rejection region, the likelihood ratio of the new outcome is smaller than the likelihood ratio of the old outcome (in this example, it would be 3/2 versus 5/2). Thus, if the alternative hypothesis is true, the probability the new outcome would occur is less than or equal to the probability that the outcome it replaced would occur. Swapping outcomes has reduced the *power* of our test. By maximizing the likelihood ratio, we can obtain the most powerful test at a given significance level.

Conclusion: We need to have both a primary hypothesis and an alternative hypothesis or alternatives firmly in mind before we begin to gather data.

To facilitate computations, the primary hypothesis may need to be reformulated as a null hypothesis. For example, suppose that X denotes the lifespan of human diploid fibroblasts raised in standard cell growth medium, and Y denotes the lifespan of human diploid fibroblasts raised in standard cell growth medium to which Vitamin E has been added. If our primary hypothesis is "The addition of Vitamin E to standard cell growth medium will increase the lifespan of human diploid fibroblasts by no less than 30 generations," we might reformulate and reverse it as "The expected value of $Y - X - 29$ is less than or equal to zero." Our one-sided alternate hypothesis would be that "The addition of Vitamin E to standard cell growth medium will increase the lifespan of human diploid fibroblasts by 30 generations or more."

One-Sided or Two-Sided?

Completing a cost-matrix similar to Table 2.3 suggests that when a comparison involves just two treatments, the alternatives will be one-sided most of the time. Is the new teaching method or the new drug more effective than the old one? Unless the improvement is of practical value, that is, unless it represents a significant increase in effectiveness, the new method will probably just fall by the wayside. When two methods or treatments are both novel, then and only then should a two-sided test be employed.

Ordered or Unordered Alternative Hypotheses?

When testing qualities (number of germinating plants, crop weight, etc.) from k samples of plants taken from soils of different composition, it is standard to use the F-ratio of the analysis of variance. For contingency tables, many routinely use the *Chi*-square test to determine if the differences among samples are significant. Both the F-ratio and the *Chi*-square are what are termed omnibus tests, designed to be sensitive to all possible alternatives. As such, they are not particularly sensitive to ordered alternatives such "as more fertilizer, more growth" or "more Vitamin E, more life extension." Tests for such ordered responses at k distinct treatment levels should properly use one of the modeling methods described in Part IV when the data are measured on a metric scale (e.g., weight of the crop). Tests for ordered responses in 2 × C contingency tables (e.g., number of germinating plants) should use the trend test described by Berger, Permutt, and Ivanova (1998).

Summary

In this chapter, we reviewed the steps in formulating hypotheses for study, in particular the steps needed to pass from qualitative objectives to quantitative testable hypotheses. A brief introduction to decision theory was provided, and you were given working definitions of null hypotheses, one- and two-sided tests, and ordered and unordered alternatives.

To Learn More

Neyman and Pearson (1933) first formulated the problem of hypothesis testing in terms of two types of error. Extensions and analyses of their approach are given by Lehmann (1986) and Mayo (1996). For more guidelines on formulating meaningful primary hypotheses, see Selike, Bayarri, and Berger (2001). Clarity in hypothesis formulation is essential; ambiguity can only yield controversy; see, for example, Kaplan (2001).

CHAPTER **3**

Coping with Variation

Prescription

- Start with your reports
- List all outcomes of interest
- Consider using surrogates
- List all sources of variation
- Describe how you will cope with each source
- Specify your collection methods
- Itemize the costs of collection
- Consider using surrogates
- Should the study be performed?

Start with Your Reports

Let your objectives determine your observations. Begin by printing out a copy of the final reports you would like to see:

> "743 patients self-administered our psyllium preparation twice a day over a three-month period. Changes in the Klozner–Murphy self-satisfaction scale over the course of treatment were compared with those of 722 patients who self-administered an equally foul-tasting but harmless preparation over the same time period.

"All patients in the study reported an increase in self-satisfaction, but the scores of those taking our preparation increased an average of 2.3 +1 0.5 points more than those in the control group.
"Adverse effects included . . ."

These reports will determine the data you need to collect. Your listing of possible outcomes, the means for measuring them, and adverse effects should reflect your review of past studies as described in Chapter 2.

Do not hesitate to write in exact numerical values for the anticipated outcomes, your best guesses. These guesstimates, for efficacy and for adverse effects, will be needed when determining sample size (see Chapters 5 and 11). Make sure you've included all end points and all anticipated side effects in your hypothetical report. Once this prototype report is fleshed out, you'll know what data you need to collect and will not waste your or your institution's time on unnecessary or redundant effort.

List All Outcomes of Interest

Here are a few guidelines:

- Objective criteria are always preferable to subjective criteria.
- True end points such as death or incidence of strokes should be employed rather than surrogate response variables such as tumor size or blood pressure.
- The fewer the endpoints, the better. A single end point is always to be preferred as it eliminates the possibility that different endpoints will point in different directions.
- On the other hand, a constellation of changes can be more diagnostic than a change in a single variable.

List All Sources of Variation

Consider the following relatively straightforward experiment: Time how long it takes to run a fixed distance on a track. Obviously, the time taken can be expected to vary from individual to individual and will depend on the distance. So suppose we record the times of 5–10 individuals over 3 track lengths (say, 50 meters, 100 meters, and a quarter mile). Since the participants (or trial subjects) are sure to complain they could have done much better if only they had been given the opportunity, record at least two times for each study subject. This results in four variables (individual, track length, trial, and time) that need to be recorded.

But surely the speed at which an individual runs will depend on his/her sex, age, race, weight, previous track experience, and current health. It's also likely to depend on the condition of the track and the material of which the track is made, the presence of spectators, and the weather (rain? wind speed and direction?). This simple experiment will produce meaningful results only if we measure, block, or control all these potential sources of variation.

Put down your thoughts in written form. Use a spreadsheet to list all possible sources of variation, as in Table 3.1.

TABLE 3.1   Time Trials: Sources of Variation

Subject	Observer	Apparatus	Environment
Sex	Sex	Length	Rain
Race	Experience	Material	Wind
Age	Motivation	Condition	Speed
Inseam			Direction
Weight			
Hemoglobin			
Experience			

Survey Questions

The fewer the number of questions you ask, the better. If your questionnaire is too long, people will not take the time to fill it out or will abandon it midway. Toward what respondents hope will be the end of lengthy face-to-face interviews, they begin to answer without thinking.

Look for ways to make each question more precise. Avoid and correct all of the following:

- Complex wording or structure. Use simple sentences and vocabulary appropriate to your audience.
- Vague or overly general questions. Ask yourself what you expect to learn from the responses— positive or negative—and you will know whether they are too vague.
- Questions that could be misinterpreted. Practically anything can. (If you've taught or taken a class, you'll know exactly what I mean.) Words can convey different meanings to different people. Conduct several pilot tests. Submit questions and request answers in writing, first from colleagues, then from strangers.
- Questions with a socially "correct" or desirable answer. A potential drawback of the face-to-face interview is that the respondent often looks to the interviewer for guidance.
- Questions that ask about more than one thing.

Make Use of Principal Component Analysis. It's important to keep questionnaires as short as possible, yet to have some seemingly redundant questions for the purpose of validating responses. Assuming you've already conducted a pilot study, principal component analysis (PCA) is a method for rapidly assessing the associations among many different questions. In effect, PCA reduces a

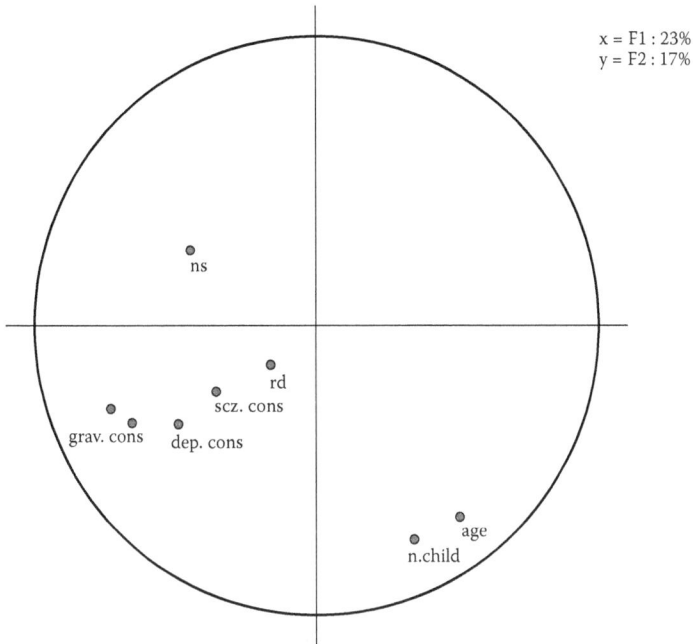

x = F1 : 23%
y = F2 : 17%

FIGURE 3.1 Relations among factors affecting prisoners' health.

hard-to-read matrix of correlations to a two-dimensional display. Here is an example, using data from a survey of prisoners' health (Figure 3.1):

```
#Install the gdata and psy packages before
  you begin
# load data supplied by Bruno Falissard
con=url(
"http://statcourse.com/research/mhp.dat")
load(con)
library(gdata)
mhp=remove.vars(mhp.ex, "prof")
library(psy)
mdspca(mhp)
```

Describe How You Will Cope with Sources of Variation

Statisticians have found four ways of coping with individual-to-individual and observer-to-observer variation:

1. *Controlling.* Making the environment for the study— the subjects, the manner in which the treatment is administered, the manner in which the observations are obtained, the apparatus used to make the measurements, and the criteria for interpretation—as uniform and homogeneous as possible.

2. *Blocking.* A clinician might stratify the population into subgroups based on such factors as age, sex, race, and the severity of the condition and restrict subsequent comparisons to individuals who belong to the same subgroup. An agronomist would want to stratify on the basis of soil composition and environment. In the present example, we have chosen to set up three blocks based on the distance to be run. We could also divide our test subjects into blocks based on their age, race, sex, and degree of previous experience.

3. *Measuring.* Some variables such as the wind speed and the condition of the track are able to take any of a broad range of values and don't lend themselves to blocking. As we show in the next chapter, statisticians have methods for correcting for the values taken by these *covariates.*

4. A fourth, purely statistical technique, in which one randomly assigns runners to different test conditions, will also be described in the next chapter.

Specify Your Collection Methods

I find it helpful to make use of a spreadsheet, as in Table 3.2. In the first column, I list (in no particular order) the names of all the variables I will be collecting. In the second, whether these will be set at some fixed value or measured, and in the third column, I specify how they will be

TABLE 3.2

▲	A	B
1	Time Trials Variables	
2	Variable	Collected by
3	Sex	Subject
4	Race	Subject
5	Age	Subject
6	Date of Birth	Subject
7	Date of Experiment	Me
8	Start Time	Me
9	End Time	Me
10	Trial #	Me
11	Hemoglobin	Clinical laboratory
12	HHematocrit	Clinical laboratory
13	Glucose	Clinical laboratory
14	ST-segment	electrocardiogram
15	troponin T	Clinical laboratory
16	barometric pressure	direct data acquision
17	room temperature	direct data acquision

collected. In a clinical trial, for example, an observation might be made by a patient, by a primary-care physician, read from a pathology slide, read from an echo-cardiogram, or received from a lab. Of course, data may also be gathered from surveys, from government records, or from other already-existing databases.

Itemize the Costs of Collection

When I'm through listing variables, both outcomes and predictors—sometimes the list seems never-ending—I sort the entries so that they are grouped by the method of collection. My rationale is that the cost of measuring blood chemistries is approximately the same whether I am measuring only the cholesterol level or the cholesterol level plus liver enzymes plus the kidney clearance rate. Similarly, the study physician's fees will usually be tied to the number of patient visits rather than to the amount of data recorded.

The costs of grouped items may be subdivided among the members of the group or assigned to the most essential of its members.

The costs of items obtained through a survey are too easily underestimated. While the costs of distributing a self-administered survey via a mass-mailing may appear rather limited, it is necessary to allow for the expense of polling nonresponders an individual at a time.

I enter the costs in three columns labeled $s, (my) time, and delays.

Consider Using Surrogates

We are now in a position to decide to move ahead measuring the outcomes and sources of variation as originally envisioned or to make use of surrogate variables in their place.

The presence of the HIV virus often serves as a surrogate for the presence of AIDs. But is HIV an appropriate surrogate? Many individuals have tested positive for HIV who do not go on to develop AIDs.* How shall we measure the progress of arteriosclerosis? By cholesterol levels? Angiography? Electrocardiogram?

Here are two other examples.

In a study of devices for maintaining flow through coronary arteries, the more accurate reading of residual stenosis would be made at high cost from an angiogram read by an independent laboratory. But letting the surgeon who performed the operation make the determination would be far less costly and entail a much shorter delay. One might be tempted to use the surgeon's guesstimate as a surrogate variable for the angiogram reading.

Consider another study in which we observe animals that have been treated for a life-threatening disease. "Survival-time" might be the experimental outcome that was visualized originally, but "time-to-incapacitation" would be a not-unreasonable surrogate.

---

* A characteristic of most surrogates is that they are not one-to-one with the gold standard.

Establish a Time Line

Will your study require before or baseline data? Almost all studies will.

Will there be only a single measurement or a series of follow-ups?

What factors will determine the end of the study?

Most important, will you be able to complete your observations in the time available? In the CARE study (Sacks et al., 1996), it took more than two years before the effects of treatment could be distinguished from that of a placebo.

The more complicated the study, the greater the importance of establishing a time line.

- Locate the participants—easy if all one has to do is order mice, hard if one has to fight to get units for testing, and most time-consuming if one has to do a preliminary screen to see if individuals satisfy the study for entry into the study.
- Make baseline measurements (see also Part II)
- Assign subjects to treatment (see Part III for methods)
- Administer treatment
- Perform initial and follow-up readings (if any)
  - Now, not while the study is in progress, is the time to establish how delays in follow-up (missed doses, missed appointments) will be recorded
- Terminate the study
  - Now, not while the study is in progress, is the time to establish how the study will be terminated.

Should the Study Be Performed?

Do you understand the mechanisms of action? If not, perhaps a preliminary study is called for.

Can you measure the desired outcome or will it be necessary to make use of a surrogate variable? If the latter, will you be able to specify the actual source of observed changes?

Should you be testing for a dose response?

Are you familiar with all the methodologies you will be employing? If not, perhaps you ought to first replicate an already published study or studies that make use of the unfamiliar methodologies.

Are there too many variables so that too many independent measurements are called for? Perhaps you ought to narrow the scope of your inquiry.

To Learn More

For more on the design of survey questionnaires, see Leung (2001), Fowler (2002), and Nardi (2006).

CHAPTER 4

Experimental Design

Prescription

It's one thing to measure shifts in a comet's position over time or the yield of a chemical process conducted in the presence of various catalysts. The major challenges and, along with them, the need for statistical methods is felt when we attempt to measure the responses of living beings.

Anyone who spends any time in a schoolroom, as a parent or as a child, becomes aware of the vast differences among individuals. My most distinct memories are of how large the girls were in the third grade (ever been beaten up by a girl?) and the trepidation I felt on the playground whenever we chose teams. In my college days, I was to discover there were many individuals capable of devouring larger quantities of alcohol than I without noticeable effect. And a few, very few others, whom I could drink under the table.

Whether or not you imbibe, I'm sure you've had the opportunity to observe the effects of alcohol on other persons. Some individuals take a single drink and their nose turns red. Others can't seem to take just one drink.

The majority of effort in experimental design is devoted to finding ways in which this variation from individual to individual won't swamp or mask the variation that results from differences in treatment. These same design techniques apply to the variation that results from one observer being extremely meticulous in her observations and another who is just passing time until retirement, or from the physician who treats one individual being more knowledgeable, more experienced, more thorough, or simply more pleasant than the physician who treats another.

In this chapter, we consider in turn the value of baseline measurements, whether one ought keep the experiment simple or gather as much information as one can, the use of blocks and strata, and the application of other, more sophisticated experimental designs.

Define the Study Population

Perform an experiment on a single mouse, and you'd best take that mouse with you to the conference so attendees can see the animal to which your research applies. One year, while teaching middle school, I had my sixth-graders measure and record their heights. Do my findings apply only to that particular class? To all the sixth graders in that school? To all sixth-graders in the county? The state? The world?

For results to be applicable to a population, they must be based on a representative random sample from that population. But the larger a population, the more potential sources of variation. Common sense dictates that your initial investigations be based on a small, narrowly defined population

The Purpose of Experimental Design

Suppose I go for a run on the track. My time t is given by the equation $\mu_{pg} + z$, where μ_{pg} is my expected running time and z is a random variable with zero expectation and variance σ^2_{wp}, where the wp stands for within person. Next, I recruit more runners. Our average time t is given by the equation $\mu + z$, where z is a random variable with zero expectation and variance $\sigma^2_{wp} + \sigma^2_{bp}$; here, bp stands for between-person variance. Finally, I conduct an experiment in which half of the runners are given aspirin before the race. Our average time t is given by the equation $\mu_A + \mu_N + z$, where z is a random variable with zero expectation and variance $\sigma^2_{wp} + \sigma^2_{bp} + \sigma^2_{bt}$; here, the bt stands for between-treatment variance. In this latter experiment, my chief interest is in the difference $\mu_A - \mu_N$. The objective of the experimental designs discussed in what follows is to minimize or eliminate the effects of the within-person and between-person variances.

The Simplest Designs

During a period of relatively unchanging weather, I run around the track at approximately the same time each day for k days in a row. Then I produce a set of summary statistics for my running times as described in Chapter 8. Note that now $t = \mu_{pg} + z'$, where z' is a random variable with zero expectation and variance σ^2_{wp}/k. Through the use of repeated trials, I have reduced the variability of my estimates, provided, that is, that the weather remains relatively unchanged and my health and energy level remain about the same.

Similarly, by increasing the number of individuals we survey, we can increase the precision of our results.

Randomized Blocks

Like Michel de Montaigne, I find myself a fascinating object of study, but journal editors and journal readers might be more interested in a study involving multiple individuals, particularly as such a study could be generalized to a still larger population. To keep my findings as precise as possible, the new participants will be divided into groups or blocks based on their sex, race, and weight/height. I will analyze the data in each block separately. In the case of surveys, this blocking can be done after the data are in hand.

If I were picking up sides for a track meet, I'd do my best to select the fastest runners. Instead, to avoid biasing my findings, I will make a random selection from passersby. This won't be easy. Those who enjoy running (generally those who are good at it) will volunteer, while those who don't will leap off their couches to get away— the most exercise they've had in years. Similar problems exist with mail-in surveys, and it is often necessary to call individually on nonresponders.

Governments have it easy; they call you to jury duty or draft you into the service, and you come. Random selection is best practiced on animals or with widgets off an assembly line. We consider methods you can use to ensure the random selection of human subjects in Chapter 5 and Chapter 10.

Nested Designs

What if the purpose of our experiments is to assess the effect of some treatment? What if we want to assess the effects of a 10-minute warm-up on running speed or of a pretrial meal restricted to pasta? Clearly, we need to conduct two sets of trials, one without the treatment and one with. Or we might perform four sets, one in which the participants skip a meal, others in which they eat what they like, eat only pasta, or eat only a small portion of pasta.

If we use a different group of subjects for each treatment, the experimental design is said to be nested. Three principles apply:

1. Control; that is, fix the environment, in so far as this is possible.
2. To forestall bias in selection, the assignment of treatments to subjects in a nested design should be made with the aid of some random mechanism (see Chapter 5). If you have blocked the study, then randomization should be performed separately and independently within each block.

3. Use at least one standard or meaningful control group. For example, in a comparison of headache remedies, the control group would be treated with aspirin, not left untreated.

In some instances, a placebo will suffice. A placebo is a treatment that from the viewpoint of the treated individual (rat or human) is indistinguishable from the real thing. Physicians and witch doctors soon learn that merely giving a person a pill (any pill) or dancing a dance often results in a cure. Of course, the pill should look and taste the same as the drug under investigation.

The researcher's attitude is as important as the treatment. If part of the dance is omitted—a failure to shake a rattle, for example; why bother if the patient is going to die anyway?—the patient may react differently.

Ideally, the individual administering the treatment will not be aware whether he is proffering a placebo or the real thing. This double blind principle also applies to experimental animals. Dogs and primates are particularly sensitive to their handlers' attitudes.

Though your focus may well be on the treatment, we advise you to use twice as many subjects in the control group than in the group or groups that receive the treatment of interest. Stuff happens. A subject comes down with the flu or stops taking their medication.

Not infrequently, the control group will consist of those receiving a standard treatment or manufactured in accordance with established practices, while the "treatment" group receives an experimental drug or one manufactured with a novel manufacturing process. The same advice applies.

Do you recall all the fuss about silicone implants? Some of the women who received these implants came down with mysterious illnesses. They sued the manufacturer—sued big—and won, and the manufacturer went bankrupt. Only much later were studies of these implants repeated, this time making use of controls. The results:

patients without implants had just as many diseases of unexplained origin as those with.

Matched Pairs

When two groups of subjects are compared in a nested design, our estimate of the treatment effect will be affected both by the within-subject and between-subject variance. In a matched pairs design, each subject acts as its own control. The subject is tested and/or measured before the treatment is administered and then tested again afterward. The resulting differences $X_{ia} - X_{ib}$ have variances $2\sigma_{ws}$, eliminating the between-subject component of the variance completely.

Case Matching

Even when it is not possible to use a subject as its own control, it still may be possible to select the experimental subjects in pairs, where the members of each pair have the same basic characteristics—age, sex, race, height, weight, hematocrit. Twins would be ideal. The treatment would be assigned at random to one member of each pair. The analysis would be conducted exactly as in the matched pair case, though a larger number of samples would be required to obtain the same precision.

Cross-Over Designs

I'm not sure that a straightforward application of matched pairs would be possible in our time trials. Would I be able to run as quickly in the second heat of the day? While if I ran the course on different days, wouldn't this introduce the possibility of changes in track conditions?

A cross-over design should be used in comparisons of treatments when each subject acts as his or her own control and there is a possibility of a carryover of the effect of one of the treatments into the time period devoted to the next. A 2×2 cross-over design would take the following form (Table 4.1).

TABLE 4.1   Plan for a Cross-Over Design

	Time period	
Treatment	**1**	**2**
A	Group a	Group b
B	Group b	Group a

Multiple Treatments and Multiple Factors

Imagine that an agronomist is interested in comparing the yields of several varieties of seed, planted on various soils, to which various levels of a fertilizer have been added. Complex experimental designs such as this lend themselves to statistical analysis provided

- The assignment of treatments to subjects (in this instance fertilizer to soil plot) is done at random
- The design is balanced, that is, equal numbers of subjects are present at each treatment combination

The complete design in which all treatment combinations are observed, and four incomplete designs—confounding, and Latin squares, fractional factorials, and incomplete blocks—are considered in what follows. A preliminary consultation with a statistician prior to the use of any incomplete design is strongly recommended.

Complete Design

A table of results (Table 4.2) says it best.

Confounding

Ideally, all our experiments ought to be carried out under approximately the same conditions. Given the effect of the environment on our runners, it would be ideal if we could carry out all the trials on the same day, and better still if we could conduct them at the same time of day. But there just may not be time. A partial solution lies in

TABLE 4.2   Observations in a Complete Design

Effect of Sunlight and Fertilizer on Crop Yield

Sunlight	Fertilizer		
	Low	Medium	High
Low	5	15	21
	10	22	29
	8	18	25
High	6	25	55
	9	32	60
	12	40	48

doing some but not all of the trials in the morning and the remainder in the afternoon. The bad news is that in doing so we confounded the circadian effects (if any) with the effects of treatment. The good news is that with a help of a statistician (who will have a book of confounded designs to help him or her) the effects of confounding will be minimized. In particular, if we let 1 denote the control group, *a* the group that receives only treatment *a*, *b* the group that receives only treatment *b*, *ab* the group that receives both treatments, and so forth, then doing 1, *ab*, *ac*, and *bc* in the morning, and *a*, *b*, *c*, and *abc* in the afternoon will minimize the effects of confounding time and treatments.

Latin Squares

We can reduce the number of samples required and, in some instances, eliminate a systematic bias by using a Latin square. Using a Latin square, we can evaluate the effects of three different factors each applied at K different levels with as few as $K*K$ observations rather than $K*K*K$.

Having a home garden, I can appreciate how radically crop yield changes from plot to plot within a space of only a few feet. Table 4.3 depicts a Latin square which has eliminated the possibility of a systematic gradient in soil quality along either a north–south or east–west gradient.

TABLE 4.3   Design for a Latin Square

	W		E
N	Hi	Med	Lo
	Lo	Hi	Med
S	Med	Lo	Hi

Table 4.3 depicts only one of 12 possible 3 × 3 Latin squares. In every instance, the choice among them should be made with the aid of a chance device.

To generate Latin square designs using R, first install the crossdes package.

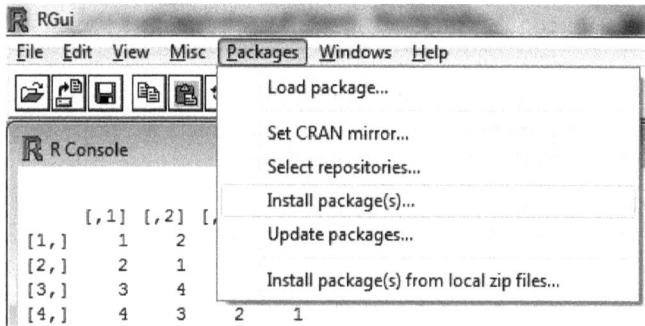

```
> library(crossdes)
```

To generate some of the possible 3 × 3 designs:

```
> MOLS(3,1)
, , 1
 [,1] [,2] [,3]
[1,] 1 2 3
[2,] 2 3 1
[3,] 3 1 2
, , 2
 [,1] [,2] [,3]
[1,] 1 2 3
[2,] 3 1 2
[3,] 2 3 1
```

If you are going to replicate your experiment, use a different design for each replicate.

Three (3) is a prime number; four (4) is not. To generate possible 4 × 4 designs:

```
> MOL(2,2)
```

As four is 2 raised to the second power.

Fractional Factorials

In a factorial experiment, the effect on some outcome of multiple factors (often three or more) each applied at two or more levels are observed. Strictly speaking, the completely randomized design (Table 4.2), blocked designs, confounding, and the Latin square (Table 4.3) are all examples of factorial designs.

Consider an investigation, perhaps one in oncology or the treatment of AIDs, in which we hope to study the effects of various combinations of four different pharmaceuticals. We will either use or withhold each drug for a total of 2 × 2 × 2 × 2 or 2^4 = 16 different combinations. Can we get by with a smaller number? (It's not just a question of the expense, but of our wish to minimize the effects on human subjects of potentially harmful combinations.) The answer is yes, provided we choose the right combinations. In particular, the half-replicate 1, *ab, ac, bc, ad, bd, cd, abcd* is recommended (where the presence of a letter indicates that the matching drug was administered).

Experiments performed by chemical engineers may well involve a dozen or more factors (including duration, temperature, pressure, and so forth) and the savings entailed in using a fractional factorial design can be quite substantial *provided* one makes use of the correct set of treatment combinations. Consult a statistician before making use of this type of design.

A statistician who wanted to generate a half-replicate of 2 × 2 × 2 × 2 design using R would first install the crossdes package as described in the previous section, and then use the following commands:

```
> dat<-gen.factorial(levels=2,nVars=4,
varNames=c("a","b","c","d"))
> optFederov(~.,dat,8)
$design
 a b c d
3 -1 1 -1 -1
4 1 1 -1 -1
5 -1 -1 1 -1
6 1 -1 1 -1
9 -1 -1 -1 1
10 1 -1 -1 1
15 -1 1 1 1
16 1 1 1 1
```

Or, using our previous notation,

b, ab, c, ac, d, ad, bcd, 1

The good news is that by using a fractional factorial design, we will have cut the number of samples required in half. The bad news is that we will have confounded the main effects of the treatments with their three-way interactions.

Main effects are defined as the effect a treatment will have when used by itself. A two-way interaction is defined in this context as the difference between the effect a treatment will have when used by itself and the effect it will have when used in conjunction with a second treatment. Used together, treatments can act independently (zero interaction), synergistically with one promoting the action of another (a positive interaction), and antagonistically (a negative interaction). Causal mechanisms for two-way interactions are readily envisioned.

A three-way interaction is defined as the differences among two-way interactions between A and B at various levels of a third factor C. k-way interactions are defined analogously. If a causal explanation for such interactions is lacking, they are very likely to be the product of chance alone. Unless such a causal explanation readily springs to mind, confounding an unlikely three- or higher-way interaction with a main effect seems a small price to pay for a reduction in sample size.

Incomplete Blocks

We make use of fractional factorial designs when the number of treatment combinations exceeds the number of plots per block. The result is that estimates of main treatment effects are aliases of higher-order interactions or other, relatively uninteresting treatment comparisons. As with any other experimental design, the assignment of treatments to specific plots within the block is made at random.

We make use of incomplete blocks when block size is severely restricted or the number of levels of one of more of the treatments is quite large. Here are two examples:

1. Five insecticides, at each of three concentrations, are to be tested in the fields. A full trial is highly laborious, and only three insecticides can be tested each day.
2. Different techniques for women in labor to self-administer analgesics are to be tested. Subject-to-subject variation in response is large, but it is impractical to ask a subject to try more than two techniques.

For R to set up incomplete blocks for the first example, with each block corresponding to a day, and two complete replications:

```
> library(crossdes)
> trt=5*3; blk=3; rep=2
> find.BIB(trt,blk,(trt*rep)/blk
```

```
[,1]  [,2]  [,3]  [,4]  [,5]  [,6]  [,7]  [,8]
[,9]  [,10]
[1,]  1  2  4  5  6  8  9  10  13  15
[2,]  3  5  6  7  10  11  12  13  14  15
[3,]  1  2  3  4  7  8  9  11  12  14
```

K.I.S.S.?

Should you keep your experiments simple or gather as much information as you can in each one? The answer depends on the context. With forms that are self-administered, the fewer the questions, the more likely the form is to be completed—and completed honestly. Have you ever hung up the telephone in the middle of a customer satisfaction survey?

The less you know, the more straightforward your experimental design should be.

In pilot experiments, limit the ranges of the variables under your control. In the time trials we considered in the previous chapter, your pilot experiment should have all subjects running the same distance on the same day and at the same time (thus controlling weather conditions—hopefully). Begin on the first day with three subjects of approximately the same age, sex, race, build, and previous track experience. Once the bugs are out—perhaps you ought to have read through the instructions for your new stopwatch before you began—consider running subjects of varying backgrounds on subsequent days.

With clinical trials and surveys, the initial runs should be viewed as tests of the adequacy of both the data collection methods and the data collection forms, and the number of subjects should be severely restricted. A somewhat larger but still restricted number of subjects should be used to estimate dropout and withdrawal rates. Again, sophisticated experimental designs are not warranted at this stage.

Human subjects necessitate simpler designs.

The use of Latin squares and fractional designs is commonplace in chemical engineering and agronomy. But complex designs need to be balanced with equal numbers in each category to be amenable to statistical analysis. Humans just cannot be relied on, and missing data due to dropouts and noncompliance are inevitable. To a limited extent this is also true with animal experiments. They get sick; they die; though, unlike humans, they are less likely to skip appointments.

Summary

In this chapter, you learned the basic principles of experimental design:

- Control the environment as far as possible.
- Measure what you can't control.
- Assign subjects to treatment using a random mechanism.
- Use the simplest possible experimental design until you feel you are in full control of the data collection methodology.

You were introduced to a number of designs, including nested designs, matched pairs, confounding, Latin squares, fractional factorials, and incomplete blocks.

To Learn More

The texts by R.A. Fisher (1925, 1935) on the basic principles of experimental design are as valid today as when they were first written, and remain among the most readable accounts on the topic.

The simplest and most straightforward text on the design of Latin squares, fractional factorials, and incomplete blocks is that of Finney (1960).

For clinical trials, Good (2006; Chapters 5 and 6) is an essential resource.

PART II
DATA COLLECTION

CHAPTER **5**

Fundamentals

Prescription

- Decide how you will make your measurements
- Prepare formal descriptions of methods and materials
- Record all measurements in a computer
- Forestall disasters

How Will You Make Your Measurements?

A good response variable

- Is easy to record—imagine weighing a live pig.
- Can be measured objectively on a generally accepted scale.
- Is measured in appropriate units.
- Takes values over a sufficiently large range that discriminates well.
- Is well defined. A patient is not « cured » but may be « discharged from hospital » or « symptom-free for a pre-defined period. »
- Has constant variance over the range used in the experiment.
- Minimizes the overall cost of data collection.

Precise measurements usually cost more to make, but require fewer observations to obtain the same precision. Here is an example:

The regression slope describing the change in systolic blood pressure (in mm Hg) per 100 mg of calcium intake is strongly influenced by the approach used to assess the amount of calcium consumed. The association is small and only marginally significant with diet histories, but large and highly significant when food frequency questionnaires are used. Diet histories assess patterns of usual intake over long periods of time and require an extensive interview with a nutritionist, whereas food frequency questionnaires are both simpler and reflect current consumption.

Collect Exact Values Whenever Possible

A second fundamental principle is also applicable to both experiments and surveys: Collect exact values whenever possible. Worry about grouping them in intervals or discrete categories later.

Formal Descriptions of Methods and Materials

At the minimum—if you are without hope of assistants—your description of methods and materials should be sufficiently comprehensive and detailed that your work can be duplicated in other laboratories by other investigators. In other words, your description can and will serve double duty as the methods and materials section of any future published reports (see Chapter 15).

If you will have additional personnel to assist you at some point in the study, then your description of methods should be sufficiently detailed so that it can serve as a step-by-step training manual. (If like me, you're bound to be summoned away from your lab on innumerable occasions to perform other unwanted and unrelated duties, your manual will provide an invaluable mnemonic.) If you think your methods are likely to change during the course of the study, this is all the more reason for you to document your original methodology, along with all changes and your reasons for making them.

Document all of the following:

- Your method of blinding
- Your method of treatment assignment

Blinding

Single blinding, in which the subject is unaware of the treatment he or she is receiving, is adopted routinely in human trials, with double blinding, in which the investigator is also unaware of the treatment, being strongly recommended.

The use of double blinding is also recommended in animal trials as control animals often fail to receive the same attention of their treated counterparts.

Triple blinding, in which one investigator administers the treatment and a second investigator records the result is strongly recommended in clinical trials. For example, a second cardiologist would read angiograms following the insertion of a stent.

Treatment Assignment

- Assign treatments independently within each block.
- Generate the assignments in groups of 8–12 subjects to avoid sequential imbalance.
- Assign before screening subjects. Subject characteristics should not be a factor `in treatment assignment.
- Ensure that allocation is concealed.
- Ensure that treatment codes are concealed.

Determine Sample Size

In order to determine the optimal sample size, all of the following must be specified:

- Smallest effect of clinical or experimental significance.
- Desired power and significance level.
- Distributions of the observables.

- Alternatives of interest.
- Statistic(s) that will be employed.
- Whether the sample size will be fixed or determined sequentially
- Anticipated losses due to nonresponders, noncompliant participants, and dropouts.

As the calculation of sample size will be determined by the statistics that will be employed, discussion of the method of calculation is deferred until Chapter 10.

Put Your Data in a Computer and Keep It There

Whether you record your observations as you go in an Excel or Open Office Calc spreadsheet or use computer-assisted data entry, you'll have three immediate advantages:

- Eliminates the need to recode and reenter data. This forestalls two potential sources of error and is accompanied by a corresponding reduction in costs.
- By tabulating and monitoring the information as it is collected, problems can be detected and corrected early on. Face it: Measuring devices can go out of tolerance; even the best-designed forms contain ambiguities, and even the best-designed trials can have unexpected consequences (also see Chapter 6).
- Ease of access. The same software that simplifies data entry makes it easy for the non–computer professional to access and display the result. It puts them in a position to immediately analyze their findings. You'll be able to spot trends, such as instruments slipping out of control—if, that is, you consult and monitor your findings as you go.

Computer-assisted data entry offers at least three other advantages:

- Immediate detection and correction of errors. Mistakes such as typographical errors and misplaced decimal points are detected and corrected at the time of entry. No data is lost as a result of lapses in memory.

- Open-ended. If inspection of the "other" category reveals that "protein imbalance" is being written in with a relatively high frequency, then "protein imbalance" can be added to the options on the pull-down menu. Printed case report forms are fixed, lifeless.

- Many regulatory agencies such as the FDA now accept and even prefer electronic submissions (also known as e-subs or CANDAs), thus doing away with the need to manage or store paper case report forms. If paper forms are required, they are readily produced. And if a paper form turns up missing, it is easily regenerated from the electronic record and submitted to the investigator for signature.

- Self-administered forms should be made available in multiple languages, much like ATMs.

Implementation of computer-assisted data entry involves five steps:

1. Linking all laboratory instruments to a central computer. (If Starbucks and McDonalds have Wi-Fi, shouldn't your lab?).
2. It can be cheaper to supply all field personnel with a minimally equipped laptop than to transcribe and misinterpret their handwritten notes.
3. Developing and testing data entry software.
4. Training personnel in the hardware's and software's use.
5. Monitoring the quality of the data.

TABLE 5.1 Data Specifications Table

Item	Group	Units	Question if	Reject unless
Year of birth	Bp	Year		17 < (Current year – Birth year) < 81
Diastolic pressure	B,Fn	mm/HG	DP < 50 or DP > 110	30 < DP < Systolic Pressure

Pre-Data Screen Development Checklist

As specified in Chapter 3, you've already grouped data items by the individual or unit that will collect them (subject, lab instrument, lab instrument, nurse, or physician) and the time at which it will be collected (initial screen, baseline, one week follow-up).

For each data item, the units and acceptable range need to be specified. See Table 5.1 for some examples.

Forestall Disaster

In your documentation, specify and provide specific examples of how the data is to be recorded. This is particularly important with dates; a good example is January 7, 1941.

For Experiments

Long ago, I designed an experiment where I had to get up every two hours over a 24-hour period, shake some petri dishes, and count the detached cells. Believe it or not, I occasionally missed a count. So, after a day or so of rest, I'd do it again. My chief competitor in the field and a good friend worked across town. He looked great when we met for lunch, while I looked beat. "Built this apparatus," he told me, "shakes the cells loose and delivers them to a counter."

- Simplify your protocol
  - Automate as much as you can
  - Reduce the number of simultaneous measurements

Ever drop a bottle?

- Inventory materials before you begin.
- Provide backups in the event of spillage.
- Verify the health of all animals.

Verify that the control preparation or process differs from the treatment only in the aspect of interest. For example, if the treatment is being administered in liquid form, it should smell and taste the same as the control, at least, in so far as a mere human can discern.

For Surveys

Decide whether you will use face-to-face interviews or self-administered questionnaires. Even should you employ the latter, face-to-face or, at the very least, telephone interviews will be required for a subsample of the nonresponders.

Web-based surveys with initial solicitation by mail (letter or postcard) and e-mail not only cut both costs and time to completion dramatically, but they also reduce the proportion of missing data and incomplete forms. Web-based surveys are easier to monitor and forms may be modified on the fly. Web-based entry also offers the possibility of displaying the individual's prior responses during follow-up surveys.

Three precautions can help ensure the success of your survey:

1. Award premiums, but only for fully completed forms.
2. Continuously tabulate and monitor submissions; don't wait to be surprised.
3. With longitudinal studies, a quarterly newsletter sent to participants will substantially increase retention (and help you keep track of address changes).

While no scientist would dream of performing an experiment without first mastering all the techniques involved, an amazing number will blunder into the execution of large-scale and costly surveys without a preliminary study of all the collateral issues a survey entails.

We know of one institute that mailed out some 20,000 questionnaires (didn't the post office just raise its rates again?) before discovering that half the addresses were in error, and that the vast majority of the remainder were being discarded unopened before prospective participants had even read the "sales pitch."

Follow these guidelines:

1. To avoid ambiguities, provide questionnaires in the native language of the respondent. Most ATMs already do this.
2. Ensure your questions don't reveal the purpose of your study, else respondents may shape their answers to what they perceive to be your needs. Contrast "how do you feel about compulsory pregnancy?" with "how do you feel about abortions?"

    What's the difference between "homosexuals" and "gay men and lesbians"? Turns out a lot—a whopping 14 percentage points of support, a New York Times-CBS News poll revealed. Only 44% of adults support the idea of "homosexuals" serving openly in the military, while 58% favor allowing "gay men and lesbians" to serve openly.
3. With populations more heterogeneous than ever before, questions that work with some ethnic groups may repulse others.
4. Be sure to include a verification question or two. In March 2000, the U.S. Census Current Population Survey added an experimental health insurance "verification" question. Anyone who did not report any type of health insurance coverage was asked an additional question about whether or not they were, in fact, uninsured. Those who reported that they were insured were then asked what type of insurance covered them.

5. If respondents are likely to fear that their responses might be used against them with questions such as, "Have you ever used methamphetamine?" and "Have you had sex with a prostitute?" use the randomized response method: Instruct respondents to flip a coin before responding, and to answer "yes" if the coin comes up heads and to tell the truth otherwise.

For Clinical Trials

Collect all clinical data in a computer, and keep it there.

Clinical trials require that physicians as well as patients be recruited and retained over a lengthy period. Keep in mind that physicians tend to overestimate the number of patients they will recruit, and some will admit patients who do not satisfy the eligibility criteria into a study in order to make their "numbers." To reduce the number of ineligibles and dropouts, pay physicians piecemeal, and be sure they are fully aware of your payment policies. Pay for completed histories and baseline values only after these forms have been reviewed for eligibility. Pay for follow-up evaluations only if these are administered in timely fashion and only after they have been scanned for completeness.

Vet physicians thoroughly before bringing them on board. Some physicians may not be good candidates because of too strong a belief in one modality or another. Others may have been guilty of misconduct or of nonadherence to protocol in a prior trial. No investigator should be brought on board without at least two interviews and a visit (not a telephone call) to the local medical society.

Provide all participating physicians with a computer, and train the appropriate staff member (usually a nurse or some other member of the physician's staff) in the use of the study software.

Emotional commitment of both patients and physicians is essential. To forestall dropouts:

- Recruit physicians and patients only *after* the design process is completed.

- Keep the interval between screening and the actual onset of treatment to a minimum.
- Maintain ongoing personal contact with the practices. Constantly endeavor to show you recognize the value of the practicing physician's time. And continue through newsletters, reports, and meetings to let the investigators and their staff know they are part of a team.
- Recognize that all face-to-face and telephone interaction with patients need be done via their study physician. Nonetheless, to encourage long-term participation of subjects, thank them as soon as they enroll via mail or e-mail, and keep them involved via regular newsletter updates.

To Learn More

Spector (2008) and Falissard (2012; Chapter 9) provide guidance on data manipulation using R.

About Clinical Trials

Good (2006: Chapters 9 and 10) reviews recruiting and retaining patients and physicians as well as computer-assisted data entry.

About Surveys

Bly (1990, 1996) shows you how to word a "sales pitch" and the optimal colors and graphics to use along with the wording. You'll learn what "hook" to use on the envelope to ensure attention to the contents and what premiums to offer to increase participation. Converse and Presser (1986), Fowler (1995), and Schroeder (1987) will assist you in wording questionnaires and in pretesting questions for ambiguity before you begin. Ostapczuk et al. (2009) and de Jong, Pieters, and Fox (2010) describe the application of the randomized response technique. John, Loewenstein, and Prelec (2012; Suppl.) studied the effects of incentives for truth-telling.

CHAPTER **6**

Quality Control

Prescription

- Be aware of potential sources of error
- Provide preventive measures
- Make baseline measurements
- Conduct a pilot study
- Monitor the data collection process
- Monitor the data

Potential Sources of Error

- Protocol deviations that result when the intervention is not performed/administered as specified
- Noncompliance of subjects with the treatment regimen
- Ineligible subjects
- Subjects who lie
- Improperly labeled formulations
- Improperly made observations
  - Inaccurate measuring devices
  - Inconsistent methods of observation, the result of
    - Ambiguous directions
    - Observer-to-observer variation
    - Time-period-to-time-period variation

- Fraud (sometimes laziness, sometimes a misguided desire to please)
- Improperly entered data
- Improperly stored data
- Missing data

Preventive Measures

- Keep the intervention simple.
- Keep the experimental design simple.
- Keep the data collected to a minimum.
- Incorporate verification questions in data entry forms.
- Pretest all questionnaires to detect ambiguities.
- Use computer-assisted data entry to eliminate transcription errors and to catch and correct data entry errors as they occur.
- Prepare highly detailed procedure manuals.

Make Baseline Measurements

Baseline measurements should be made before assigning treatments to subjects, not only to permit before/after comparisons on a subject-by-subject basis but for all of the following reasons:

- To identify subjects who for one reason or another are ineligible to participate in the study. For example, elevated white cell counts and higher-than-normal body temperatures reveal subjects who are too sick to participate.
- To compare subjects with higher than normal scores against those with lower than normal. Further blocking prior to treatment assignment may be warranted. (A comparison of the topmost quartile with the members of the lowest would be even more revealing.)
- If the factors that divide the high and low scoring are not evident from an examination of the data in

hand, additional data per subject beyond that originally visualized will be needed for a distinction to
be made.

- The study protocol should be modified.
- To provide preliminary tests of the software, hardware, forms, and questionnaires that will be used
later on.

Conduct a Pilot Study

Include 9–10 study subjects or units in the pilot study.
Include both control and treatment subjects. If multiple
treatments are to be employed in the final study, use the
intervention with the highest risk in the pilot.

Faithfully follow the study protocol. Don't skimp on the
details. If assistants are to be employed in the final study,
use at least one assistant (the least helpful) in the pilot.

Verify that the control treatment is (almost) indistinguishable from the treatment of interest. Drugs and
foods used in an investigation should have a similar taste
and appearance. An experiment cannot be described as
"blinded" if subjects can distinguish among treatments.

Monitor the study as completely as you would the
final study. This will also serve to test the adequacy of
your monitoring.

Revise the number of subjects to be used in the final
study in the light of newly acquired information on sample variability and proportions of missing data.

If your evaluation of the pilot study results in substantial changes to the protocol, consider performing a second
pilot study.

Monitor the Data Collection Process

Do the subjects understand the questions? Do they answer
truthfully? A subsample of those completing printed and
online surveys should be contacted for personal interviews to verify their responses.

Monitor the Data

- Are the baseline values of the various treatment groups comparable?
- Are the controls appropriate? Find out why not.
- Do any of the recorded data lie outside normal boundaries? The sooner "flawed" responses are investigated, the greater the likelihood that the correct values can be established.
- Are forms being returned only partially complete? Is this the result of specific questions constantly going unanswered? Resolve problems by rewording questions, breaking them into multiple parts, or utilizing the randomized-response method.

The pattern of missing data may be diagnostic, calling for further investigation. For example, the addicted might deliberately avoid answering certain questions for fear of being caught out. The R function `naclus` in the `Hmisc` library can help in detecting clusters of missing data.

```
> plot(naclus(suspect.data)
```

If such clusters contain key variables that you already know are associated with the effects of interest, a subsample of the nonresponders may be necessary. This is another reason why monitoring should be ongoing and not delayed until all responses are in hand.

To Learn More

Consult the excellent documents available from the United States Environmental Protection Agency at http://www.epa.gov/quality/dqa.html. See also, Husted et al. (2000). For other methods, see Good and Hardin (2009, Chapter 4). For insight into detecting clusters of missing data, see Falissard (2012; pp. 160–165).

PART III

ANALYZING YOUR DATA

Prescription

Get a clear picture of your data through summary statistics and graphs before you consider formulating and testing hypotheses.

CHAPTER 7

Describing the Data

Accurate, reliable estimates are essential for effective decision making. In this chapter, we compare various potential estimates of central tendency and dispersion and describe a variety of methods you may use to obtain them.

Prescription

1. Select descriptive statistics that have all or almost all of the following properties: unbiased, consistent, minimum loss, efficient, and robust.
2. Do not be concerned with computational complexity, not with a computer at hand.
3. Use both text and graphics to describe your data. But do not use graphics that require a thousand words to explain.
4. Use confidence intervals in preference to point estimates.

Box-and-Whiskers Plot

A box and whiskers plot of your observations on a single variable obtained with R's box plot command similar to that shown in Figure 7.1 is easily worth a 1000 words, that is, once the viewer fully understands what the plot is telling him or her. I shall try to take less than 1000 words to explain.

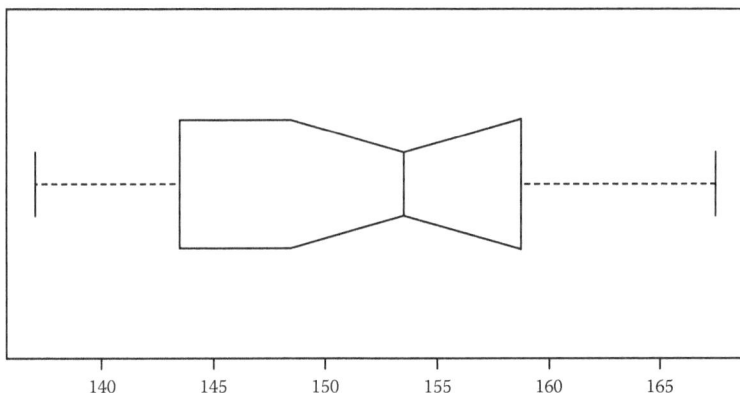

FIGURE 7.1 Box plot of heights of sixth-graders (in cms).

The whiskers that extend out on each side of the plot mark the extremes, the minimum and the maximum of the sample. Unfortunately, the minimum and the maximum are the least reliable of the estimates of the population characteristics that may be derived from the sample, as the extremes for the entire population may be quite a bit smaller or larger, respectively, than those of the sample.

The box covers the middle 50% of the sample, extending from its 25th percentile to its 75th. The notch in the middle of the box marks the arithmetic mean of the sample. Note in this instance, that the box is skewed toward smaller values, so that the center of the population, the median or 50th percentile, is smaller than the arithmetic mean.[*]

A box and whiskers plot does not tell us the number of modes the population has, that is, whether it constitutes a single population or is a mixture of several populations. Figure 7.1 displays the heights of a sixth-grade class I once taught; it is a mixture of two populations, boys, who tend to be smaller at that age, and girls, who tend to be taller.

It also does not tell us what we ought to be reporting, the plot itself (along with a lengthy explanation),

---

[*] Recall that the kth percentile is defined as the value such that a kth percentage of the population takes equal or smaller values and $(100-k)$ percentage take equal or larger values.

individual values, or summary values. This is the topic of the next few sections.

Which Statistic?

We often have a choice of several *statistics*, that is, functions of the data at hand, with which to describe a sample. Normally, we select a statistic which will provide some insight into the characteristics of the larger population from which our sample is drawn. The proper starting point for the selection of the "best" statistic is with the objectives of our study: If our statistic takes the value $S(x_1, x_2, ..., x_n)$, where x_i denotes the ith observation, and the actual value of the unknown population parameter that we wish to estimate is θ, what losses will we be subject to?

The majority of losses will be monotone nondecreasing in nature, that is, the further apart the estimate S and the true value θ, the larger our losses are likely to be. Typical forms of the loss function are the absolute deviation $|S-\theta|$, the square deviation $(S-\theta)^2$, and the jump, that is, no loss if $|S-\theta| < \delta$, and a big loss otherwise. Or the loss function may resemble the square deviation, but take the form of a step function increasing in discrete increments.

Desirable estimators share the following properties:

- Impartial: It doesn't matter in what order you examine the observations or what units the measurements are in.
- Unbiased: The expected value of the sample statistic, that is, the expected average of the statistic over many samples, is the value of the population characteristic we are trying to estimate.
- Consistent: The larger the sample, the greater the probability that the sample statistic will be close in value to the corresponding population characteristic.
- Minimum loss: Of all possible estimators, S minimizes the expected loss $L(S-\theta)$.

- Efficient: Of all possible estimators, of θ, *S* requires the least number of observations to reduce the potential loss below some fixed value *L*.
- Robust: The efficiency of the estimator does not depend on the form of the distribution of values in the population.

Center of the Distribution

Most journal editors and reviewers expect to see the arithmetic mean or average of the values of the observations in your reports, in symbols, $\sum_{i=1}^{n} X_i/n$, where *n* denotes the sample size and X_i denotes the *i*th observation. The arithmetic mean has all of the following properties:

- Impartial: It doesn't matter in what order you sum the data or what units the measurements are in.
- Unbiased: The expected value of the sample mean is the population mean.
- Consistent: The larger the sample, the greater the probability that the sample mean will be close in value to the population mean.
- Minimum squared loss: Of all possible estimators of the population mean, if the population variance is bounded,[*] using the sample mean minimizes the expected value of the square of the difference between the estimate and the population mean.
- If *X* contains the vector of your observations, mean(*X*) yields the mean of *X*.

But the arithmetic mean also has substantial limitations:

- One or two extreme values, the result of a misplaced decimal point, for example, can completely distort its value.

---

[*] If you find it hard to visualize something that isn't bounded in value, such as an earthquake, a flood, a tornado, or a wave, don't even think about designing nuclear reactors, dams, levees, bridges, or tornado-proof, earthquake-proof school buildings.

- Its use is inappropriate when the measurement scale represents relative rather than actual values. Consider the A, B, C, and D ratings a subject might assign to a movie or a video game. We can't add up or average such relative ratings.
- When a population is highly skewed, as with the population shown in Figure 7.1, the arithmetic mean can be misleading.

The arithmetic mean corresponds to a balance point. A playground see-saw requires several smaller children to balance the weight of a single overweight child. The presence of a few millionaires in their multimillionaire dollar homes can far outweigh the far larger numbers of the homeless and the under- and unemployed, at least as far as the mean income and mean home price in a neighborhood are concerned.

The *median* or 50th percentile of a population can be determined even for values measured on a relative or ordinal scale and remains unaffected by the presence of a few very large or very small outlying values. The median, too, is impartial and consistent, and it minimizes the expected value of the absolute value of the deviation between the estimate and the population median. Moreover, when a population is symmetric about its mean as in Figure 7.2 (a normal or Gaussian distribution), the mean and the median coincide.

If X contains the vector of your observations, median(X) yields the median of X.

The Hodges–Lehman estimator, once strictly an academic curiosity, may be preferable due to its efficiency and robustness for describing the center of a distribution of continuous measurements than either the mean or the median. This estimator is defined as the median of the n(n–1) pairwise averages

$$\hat{} = \mathrm{median}_{i \le j} \left(X_j + X_i \right) / 2$$

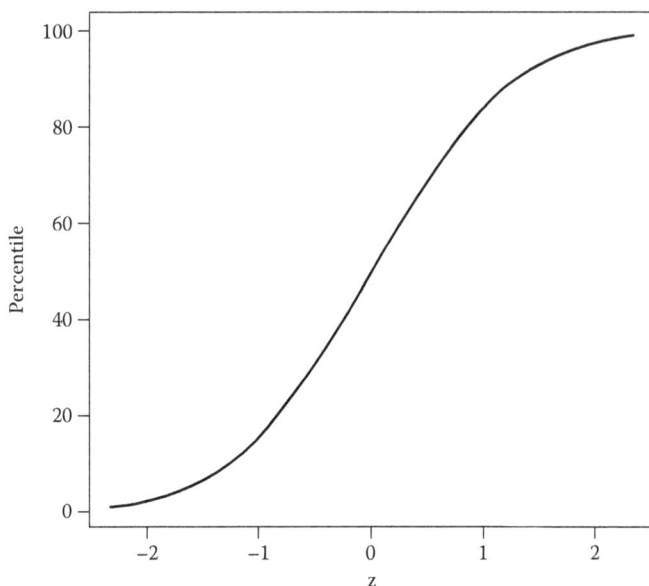

FIGURE 7.2 Cumulative frequency distribution of a normally distributed random variable (z in standard deviations about the mean).

This formula is far too complicated if one is calculating its value by hand. But who does hand calculations these days? A program in R for its computation is provided in a sidebar.

Sidebar: Computing the Hodges–Lehman Estimator

```
L=length (Data)
Data=Data/2
space=numeric((1+L)/2)
cnt=1
for (j in 1:L)
  for (i in 1:j){
    space[cnt]=Data[i]+Data[j]
    cnt=cnt+1
    }
median(space)
```

TABLE 7.1   Relative Sample Sizes Required to Obtain the Same Precision

	Distribution	
	Normal	Cauchy
Median	100	100
Mean	64	∞
Hodges–Lehmann	66	100

The advantage of this estimator, as can be seen from Table 7.1, is that it requires almost the same number of observations as the better of the mean or median regardless of how the observations are distributed.

Dispersion

Recall that our primary motivation for using statistics is to compensate for the variation in the outcomes we are observing. An estimate of the amount of variation in our data is essential.

Two obvious measures of dispersion, both provided by the box plot of Figure 7.1, are the *range* that extends from the minimum to the maximum of the sample between the whiskers, and the *interquartile range*, the difference between the 75th and 25th percentiles of the distribution (that is, the width of the box).

All the percentiles of a distribution can be read from a graph of the cumulative distribution such as Figure 7.2, which depicts the cumulative distribution of a normal or Gaussian distribution. A glance at this chart reveals not only the median (the 50th percentile), zero in this instance, but the 10th and 90th percentiles as well. These last would be of particular value to government agencies concerned with the health and welfare of a population.

Another common measure of dispersion is the sample *variance*, defined as the average of the squared deviations of the observations about the sample mean. In symbols, this would be $\sum_{i=1}^{n}(x_i - \bar{x})^2 / (n-1)$. Oops,

TABLE 7.2 Dimensional Analysis

Observations	inches, pounds, miles/gallon
Mean, Median, Percentiles	inches, pounds, miles/gallon
Variance	inches^2, pounds^2, (miles/gallon)^2
Standard Deviation	inches, pounds, miles/gallon

my bad; this last would be a biased estimate. The deviations from the mean are linearly dependent, that is, if we know $n - 1$ of them, we can easily calculate the nth. To obtain an unbiased estimate, we need to use the formula $\sigma^2 \sum_{i=1}^{n} (x_i - \bar{x})^2 / (n-1)$. Use var($X$) in R to obtain the variance.

As Table 7.2 reveals, the square root of the variance, termed the *standard deviation*, sd(X), is of greater practical value than the variance as it is measured in the same units as the observations. The standard deviation of the sample is both an unbiased estimate of the standard deviation of the population and a consistent one.

If our observations vary among themselves, then so, too, must the estimates (or statistics) based on these observations including the sample mean and sample standard deviation. Thus, any point estimate is guaranteed to be inaccurate to some degree. Even the vaunted maximum likelihood estimate has a vanishingly small probability of being correct.* We can get around this problem by developing interval estimates.

Interval Estimates

We may use a cumulative distribution function similar to that shown in Figure 7.2 to find a confidence interval for the population mean of any random variable that has an almost normal distribution.

---

* Indeed, maximum likelihood estimates have desirable properties such as unbiasedness and minimum variance only if the data is drawn from a normal distribution. For that matter, minimum variance is desirable only if our losses from errors in estimation are proportional to the square of the errors.

Suppose we are looking for a confidence interval that will include the true value of the population mean 80% of the time. The lower bound of this confidence interval corresponds to the 10th percentile of the distribution, and the upper bound to the 90th percentile.

The X-axis of Figure 7.2 is divided into units of the standard deviation of the mean, commonly referred to as the standard error of the mean, given by the formula $\sigma/\sqrt{n}$, where σ is the population standard deviation. If we know σ, then we can obtain the confidence interval directly from Figure 7.2. But it is far more likely that we will need to estimate σ from the standard deviation of the sample. In which case, we will need the use the cumulative distribution of a t-distribution with $n - 1$ degrees of freedom.

As an example, suppose we have a sample X of 16 measurements whose mean is 28 miles per gallon and whose standard deviation is 4 miles per gallon. The standard error of the mean would be 1 mile per gallon obtained from sd(X)/sqrt(length(X)). Using the R function qt(c(.10,.90),15), we can state with 80% confidence that the true value of the population mean lies somewhere in the interval from 28 – 1.34 to 28 + 1.34 miles per gallon.

Does this mean that the true value of the population mean lies in the center of this confidence interval? No. Only that with probability 80% it lies somewhere in that interval, and it may lie outside the interval (100 – 80)%, or 20% of the time. This is because the bounds of each confidence interval depend on the sample used to derive them; thus, the bounds will vary from sample to sample.

A central limit theorem tells us that a variable will have the normal distribution if it is the sum of a large number of variables, each of which makes only a small contribution to the total. As the mean of a sample is the sum of all the observations in the sample divided by the number of observations, in a sample of size n, each observation contributes only $1/n$th to the mean. Thus, when n is large, the distribution of the sample mean will be close to that of a normal distribution.

Confidence Intervals for the Population Mean

But how large does a sample have to be? If the distribution of the values of the observations in the population from which they are drawn is almost symmetric and most of the values are clustered closely about the mean, then a sample size of six is sufficient. But if the distribution is severely skewed, as would be the case if we'd measured the number of successes in, say, 20 trials for each of which the probability of success was quite low, then we would need at least 20–30 repetitions of the 20 trials to make use of this approximation.

To be fair, we'd be better off in the preceding example having the computer look up a confidence interval for us in tables of the binomial distribution. In fact, if we know our observations come from some well-tabulated distribution, for example, times to failure often come from an exponential or a *Chi*-square distribution, then we would have our computer software obtain a confidence interval for the mean time to failure from the corresponding distribution.

Any method of estimation, whether a point or interval estimate, must be appropriate to the distribution of the data that is to be estimated. A frequent error in the astrophysical literature is to apply methods appropriate to a continuous distribution—such as the normal or multivariate normal distribution—to discrete data that consists of counts that may well have a Poisson distribution.

When data is drawn from a mixture of several different distributions (as is the case with data derived from both men and women), the data must be divided into two or more strata prior to being analyzed. Of course, the strata should also be appropriate for the data in hand.

I'm often given access as an expert witness to data arising from audits of submissions by Medicare practitioners. The distribution of one such sample is depicted in Figure 7.3. It can be seen that the sample divides into two populations, those without errors and those with. An appropriate method of estimation would consist of two stages: In the first stage, an attempt would be made to obtain a lower confidence bound for the proportion of

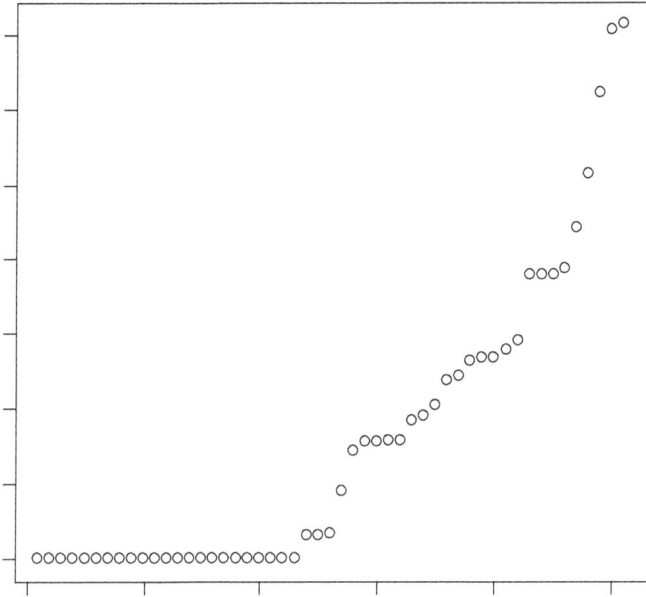

FIGURE 7.3 Medicare overpayments in dollars.

errors. In the second stage, a lower confidence bound for the expected value of an error would be obtained.

Confidence Intervals for Proportions Estimated from Randomized Responses

Proceed in stages. Suppose we observe that the proportion of "no" responses out of 400 replies to the question, "Have you used methamphetamine?" is 40%.

```
qbinom(c(.05,.95),400,.4)/400
[1]  0.36 0.44
```

But 50% of the "yes" response resulted from the flip of a coin, so our 90% confidence interval for the true proportion of "no's") is

```
2* qbinom(c(.05,.95),400,.4)/400 =
  (0.72, 0.88)
```

Confidence Intervals for Other Population Characteristics

In this section, we consider several alternate methods for obtaining confidence intervals for characteristics of the population other than the mean, such as the median and the variance.

Large Sample Methods. As the sample size increases, the sample median will come closer and closer in value to the population median. As with the sample mean, the rate of convergence is proportional to the square root of the sample size (Lehman, 1999, p. 80). With larger samples, the distribution of the sample variance will come closer to that of a *Chi*-square distribution with $n - 1$ degrees of freedom; as a result, we may use this distribution to obtain a confidence interval for the population variance.

Using a Hypothesis Test to Obtain a Confidence Interval. Let $A(\theta^*, \alpha)$ denote the acceptance region of a two-sided test of the primary hypothesis that θ is equal to θ^* against the alternative hypothesis that θ is not equal to θ^*, such that the probability of making a Type I error is less than α. That is, $S(x_1, x_2, ..., x_n)$ belongs to $A(\theta^*, \alpha)$ only if both the following conditions are met:

i. We accept the hypothesis $\theta = \theta^*$
ii. If $\theta = \theta^*$, the probability that S does not belong to $A(\theta^*, \alpha)$ is less than α.

A $1 - \alpha$ confidence interval for θ based on S, call it $C(\theta, 1 - \alpha)$, consists of all the values of θ for which S belongs to the acceptance region $A(\theta, \alpha)$. Even with small samples, we may use a distribution-free test to obtain confidence intervals for sample characteristics. We'll provide several examples of this in the next chapter, first deriving a test that does not depend on how the data is distributed, and then using this test to obtain a confidence interval.

The Bootstrap. The bootstrap may be used to obtain confidence intervals for population characteristics when no other method is appropriate. To bootstrap, we treat the original sample of values as a stand-in for the population and resample from it repeatedly, with replacement, computing the desired estimate each time.

Consider the following set of 22 observations ordered from smallest to largest. Note that the median of the sample, a plug-in estimate of the median of the population from which it is drawn, is 153.5.

137.0 138.5 140.0 141.0 142.0 143.5 145.0 147.0 148.5
150.0 153.0 154.0 155.0 156.5 157.0 158.0 158.5 159.0
160.5 161.0 162.0 167.5

Suppose we record each observation on an index card, with 22 index cards in all. We put the cards in a big hat, shake them up, pull one out, and make a note of the value recorded on it. We return the card to the hat and repeat the procedure for a total of 22 times until I have a second sample that is the same size as the original. Note that we may draw a specific card several times as a result of using this method of sampling with replacement.

As an example, our first bootstrap sample, which I've arranged in increasing order of magnitude for ease in reading, might look like this:

138.5 138.5 140.0 141.0 141.0 143.5 145.0 147.0 148.5
150.0 153.0 154.0 155.0 156.5 157.0 158.5 159.0 159.0
159.0 160.5 161.0 162.0

Several of the values have been repeated as we are sampling with replacement. The minimum of this bootstrap sample is 138.5, higher than the minimum of the original sample; the maximum at 162.0 is less, while the median remains unchanged at 153.5.

137.0 138.5 138.5 141.0 141.0 142.0 143.5 145.0 145.0
147.0 148.5 148.5 150.0 150.0 153.0 155.0 158.0 158.5
160.5 160.5 161.0 167.5

In this second bootstrap sample, we again find repeated values; this time the minimum, maximum, and median are 137.0, 167.5, and 148.5, respectively.

Rather than repeatedly drawing from a hat, a practical procedure would be to use the following R commands:

```
#This program selects 100 bootstrap
  samples from the classroom data, computes
#the variance of each sample and then
  produces a boxplot and stripchart of the
  results.
class=c(141,156.5,162,159,157,143.5,154,
  158,140,142,150,148.5,138.5,161,153,145,
  147,158.5,160.5,167.5,155,137)
#set number of bootstrap samples
N =100
stat = numeric(N)  #create a vector in
  which to store the results
                   #the elements of the
  vector will be numbered from 1 to N
#Set up a loop to generate a series of
  bootstrap samples
for (i in 1:N){
     #bootstrap sample counterparts to
     observed samples are denoted with "B"
     classB= sample (class, replace=T)
     stat[i] = var(classB)
     }
boxplot (stat)
stripchart(stat)
```

The bootstrap can be used to determine the precision of any descriptive statistic. For example, the variance of our sample is 76.7. The variances of 100 bootstrap samples drawn from our sample range between 47.4 and 115.6 with a mean of 71.4. These values provide a feel for what might have been had we sampled repeatedly 100 times from the original population. The values of the sample variances from our 100 bootstrap samples are summarized in Figure 7.4.

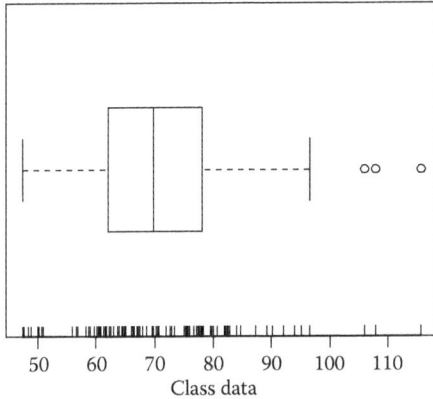

FIGURE 7.4   Box plot and strip chart of variances of 100 bootstrap samples.

An Improved Bootstrap

Alas, the confidence intervals derived from this primitive bootstrap are biased, inaccurate, and wider than they ought to be except when the observations come from a near-normal distribution, that is, when we don't need to bootstrap. By making use of this fact, the bias-corrected BC interval due to Efron and Tibshirani (1986) represents a substantial improvement.

Suppose θ is the parameter we are trying to estimate, $\hat{\theta}$ is the estimate, and we are able to come up with a monotone increasing transformation t such that $t(\hat{\theta})$ is normally distributed about $t(\theta)$. We could use this normal distribution to obtain an unbiased confidence interval for $t(\theta)$, and then apply a back-transformation to obtain an almost unbiased confidence interval for θ.

The method is not as complicated as it reads because we don't actually have to go through all these steps, merely agree that we could if we needed to. The resulting formula is already incorporated in several computer programs. In R, we would proceed as follows:

1. Download the boot library
2. Construct an R function to generate and hold the value of an existing R statistics functions such as `median()` or `var()` whose value you want a bootstrap interval estimate of. For example,

```
f.median<- function(y,id){
+ median( y[id])
+ }
```

where R knows id will be a vector of form 1:n.
3. Apply the `boot.ci()` function as in the following example, which calculates a 90% confidence interval for the median of the data based on 400 bootstrap samples.

```
boot.ci(boot(data, f.median, 400),
  conf = 0.90)
```

While I recommend the bootstrap to use in obtaining confidence intervals for characteristics of the population that rely primarily on data from the center of the distribution such as the mean, median, and 40th through 60th percentiles, it's not applicable for characteristics that rely heavily on data from the tails, such as the variance, the minimum, maximum, and the 10th and 90th percentiles.

As we will learn in the following chapters, the bootstrap is also valuable for model validation, and testing hypotheses when all else fails.

Summary

In this chapter, you were introduced to the desirable properties of estimators, to estimates of the central value, dispersion and percentiles of a population, to confidence intervals and methods for deriving them, to the bootstrap, and to the box plot and strip chart

To Learn More

Selecting more informative end points for confidence intervals is the focus of Berger (2002) and Bland and Altman (1995).

Lehmann and Casella (1998) provide a detailed theory of point estimation.

Robust estimators are considered in Huber (1981), Maritz (1996), and Bickel et al. (1993).

Additional examples of bootstrap estimation procedures may be found in Efron and Tibshirani (1993). Chernick and LaBuddde (2011) provide extensive R code for bootstrap applications. So, too, does Good (2011).

Shao and Tu (1995; Section 4.4) provide an extensive review of bootstrap estimation methods along with a summary of empirical comparisons. Potential flaws in the bootstrap approach are considered by Schenker (1985), Diciccio and Romano (1988), and Efron (1988, 1992). Canty et al. (2006) provide a set of diagnostics for detecting and dealing with potential error sources.

Carroll and Ruppert (2000) show how to account for differences in variances between populations.

CHAPTER **8**

Hypothesis Tests

Prescription

Estimation and hypothesis tests may both be viewed as part of broader category: decision theory. We wish to estimate or test a hypothesis concerning the value of some population parameter (the population mean, its variance, or some percentile) θ. We use a sample statistic $\theta^*[x_1, x_2, \ldots x_n]$ to make the decision and suffer a loss $L[\theta - \theta^*]$, where L is a nondecreasing function of the difference between our decision θ^* and the true value θ of the parameter.

After gathering data and calculating a statistic, modern methods of hypothesis testing limit us to three different decisions:

- Our hypothesis is correct; we accept the primary hypothesis.
- We can't tell whether our primary hypothesis is true or not; we are unable to reach a conclusion based solely on the data in hand.
- The primary hypothesis is not correct; we reject it and accept the alternate hypothesis.

The tests described in this chapter and the next have the following properties:

■ They are unbiased; that is, we are more likely to accept the primary hypothesis if it is true than if it is false.

■ They are exact or almost so; that is, the significance level (the probability that in repeated trials we will reject the primary hypothesis when it is true) is exact and not an approximation.

■ The power of the test, that is, the probability that in repeated trials we will reject the primary hypothesis when an alternate hypothesis is true, is equal to or greater than that of any other unbiased test.

The appropriate statistical method depends on seven factors:

1. The number of treatments (or treatment combinations) we wish to analyze or compare.
2. Whether the observations for any given treatment are exchangeable, that is, whether the order in which we observe, record, or analyze them will not affect the result.
3. Whether the observations for any given treatment are identically distributed.
4. How the parameters we wish to estimate are distributed.
5. The nature of the alternative hypotheses, one-sided or two-sided, ordered or unordered.
6. Whether there is a single outcome or multiple outcomes.
7. The type of data we wish to analyze.

This chapter is limited to the analyses for a single (univariate) outcome of observations that are exchangeable and identically distributed. We'll consider other possibilities in the next chapter.

Types of Data

The statistics we use will depend on the type of data we have gathered:

- Categorical. The data falls into mutually exclusive categories, for example, males versus females.
- Ordered (non-additive) categories. Satisfaction scales (very satisfied down to very dissatisfied) and levels of fertilizer (High, Medium, and Low) are examples.
- Counts such as the number of successes in n trials, or the number of super-clusters of galaxies in a given volume of space.
- Time between counts (or events).
- All other continuous measurements such as height, weight, pH, luminosity, and so forth.

Analyzing Data from a Single Population
Binomial Trials

If a trial can end only in success or failure, the probability p of success is the same for all trials, and the outcomes in successive trials are independent of one another, then the number of successes in n trials will have a binomial distribution.

Suppose we wish to test that the coin our friend is flipping is fair, that is $p = 0.5$. We decide to flip the coin 12 times and count the number of heads.

```
> qbinom(.95,12,.5)
[1]  9
```

that is, 95% of the time, the number of heads should be less than or equal to 9. To be precise

```
> pbinom(9,12,.5)
[1]  0.9807129
```

more than 98% of the time, the number of heads expected in 12 trials with a coin whose probability of success is 0.5 should be less than or equal to 9.

Of course, a fair coin also should not give rise to an excessive number of tails.

```
> qbinom(.95,12,.5, lower.tail=F)
[1] 3
```

In a two-sided or two-tailed test, we also would reject the hypothesis that our friend's coin was a fair one if the number of heads were 3 or less.

Inverse Binomial Sampling

Sometimes, it proves more convenient not to specify the number of trials in advance but to keep sampling until a specified number m of successes have been observed. Again, we assume that the probability p of success is the same for all trials, and the outcomes in successive trials are independent of one another.

T, the number of trials required to get m successes has the negative binomial distribution. Let's track progress at the local high school in terms of the number of unexcused absences a student has. The principal assures us that unexcused absences are rare events, no more than one or two a month. We plan to check the attendance rolls until six unexcused absences are observed. Assuming 22 days of school each month, and 1.5/22 "successes" each month, this study should take us about

```
> qnbinom(.5,size=6, prob=1.5/22)
[1] 77 days
```

but could easily take as long as

```
> qnbinom(.95,size=6, prob=1.5/22)
[1] 146 days.
```

Poisson Data

The number of events in a given interval of time or space will have a Poisson distribution providing that events in nonoverlapping intervals take place independently of one another. For example, Neyman and Scott (1953) reported that the space-time distribution of super-clusters of galaxies fit a Poisson distribution.

Dividing the data from catalogue scl_cat_ls00180_e.dat to be found at http://atmos.physic.ut.ee/~juhan/super/super_lrg/ into 9600 equal-sized regions, yields the following matrix:

```
> data
      freq  count
[1,] 7178      0
[2,] 2012      1
[3,]  370      2
[4,]   38      3
[5,]    1      4
[6,]    1      5
```

After downloading and installing the *vcd* library of R functions:

```
> library(vcd)
> goodfit(data,type= "poisson",method= "ML")
  Observed and fitted values for poisson
  distribution with parameters estimated
  by 'Maximum Likelihood'
```

count	observed	fitted
0	7178	7115.5599744
1	2012	2130.9619715
2	370	319.0893577
3	38	31.8535383
4	1	2.3848678
5	1	0.1428436

Suppose that we wish to test the hypothesis that lambda, the probability that a super-cluster will arise in one of the sum(freq) = 9600 regions, is 0.3 or greater.

Begin by putting the matrix into a data frame and then attaching it so that we can make use of the variables "freq" and "count" directly.

```
> data.frame (data)
> attach (data)
```

To determine whether lambda ≥ 0.32, we would want to reject if the total number of observed super-clusters is *less than* a specific cutoff value.

```
> qpois(.05,0.32*sum(freq))
2981
```

The total count is `sum(freq*count)` = 2875. Consequently, we reject the hypothesis that lamda≥ 0.32 at the 5% significance level.

Time between Counts

Time-to-event or time-between-counts data arises in the analysis of the time to recovery or time to death in medical studies and in engineering studies of mean time to failure. As the trials are often stopped at some prefixed time and some subjects in survival analysis may be lost to follow-up, the data may be incomplete or censored and special techniques are called for. Type I censoring in which all subjects are studied for the same period of time is considered in Chapter 10. Type II censoring in which subjects may exit the trials after a random period of time is considered in Chapter 13 under the headings of "Logistic Regression" and "Survival Analysis."

Measurements

Recall from Chapter 8 the data I'd gathered on the heights of the sixth-graders: class = c(141,156.5,162,159, 157,143.5,154,158,140,142,150,148.5,138.5,161,153,145, 147,158.5,160.5,167.5,155,137)

To test the hypothesis that the mean height of sixth-graders is less than or equal to 150 cm, we make use of Student's *t*-test, just as statisticians have been doing for almost a century.

```
> t.test(class, alternative="greater",
  mu=150)
  One Sample t-test
data: class
t = 0.8399, df = 21, p-value = 0.2052
alternative hypothesis: true mean is
greater than 150
95 percent confidence interval:
 148.3553 Infinity
sample estimates:
mean of x 151.5682
```

I've underlined the most relevant parts of R's output. The *p*-value, the probability that the observed value of would be greater than or equal to the value we observed, is 0.20 or 20% (those extra digits are meaningless and should never be reported). The one-sided confidence interval tells us that with probability of 95% the mean height might be as low as 148.4 cm.

Comparing Two Populations
Categorical Data

Does a new drug offer significant advantages over the existing therapy? Table 8.1 records the results of a therapeutic intervention with the two drugs. Let p_N denote the probability that a patient will improve given the new drug. Let p_o denote the probability that a patient given the old drug will improve. The *odds ratio* is defined as $\dfrac{p_N(1-p_o)}{(1-p_N)p_o}$. If the efficacy of the two drugs is the same, then the odds ratio will be 1. If the new drug represents an improvement, then the odds ratio will be greater than 1.

```
> data=c(9,1,5,5)
> data=dim(2,2)
> fisher.test(data, alternative="g")
```

TABLE 8.1 Results of a Therapeutic Intervention

	Drug A	Drug B
Response	5	9
No Response	5	1

Odds ratio is (9/1)/(5/5).

```
Fisher's Exact Test for Count Data
data: data
p-value = 0.07043
alternative hypothesis: true odds ratio is
 greater than 1
95 percent confidence interval:
 0.8532592 Inf
sample estimates: odds ratio 8.018207
```

Note that the confidence interval includes 1, which means that given the present small sample size, we are unable to rule out the possibility that the two drugs have the same efficacy.

For example, if we'd observed similar frequencies after taking twice as many observations, that is,

```
> datax2
     [,1] [,2]
[1,]  18   10
[2,]   2   10
> fisher.test(datax2, alternative="g")
        Fisher's Exact Test for Count Data
   data:  datax2
   p-value = 0.006907
   alternative hypothesis: true odds ratio
    is greater than 1
   95 percent confidence interval:
    1.754157      Inf
   sample estimates: odds ratio 8.489517
```

If it has been a long time since your last statistics class or, worse, if it's been an even longer time since

TABLE 8.2   Oral Lesions in Three Regions of India

Site of Lesion	Kerala	Gujarat	Andhra Pradesh
Labial Mucosa	0	1	0
Buccal Mucosa	8	1	8
Commissure	0	1	0
Gingiva	0	1	0
Hard Palate	0	1	0
Soft Palate	0	1	0
Tongue	0	1	0
Floor of Mouth	1	0	1
Alveolar Ridge	1	0	1

your instructor's, then you may have considered using the *Chi*-square approximation. Don't. It's just that, an approximation, and *Chi*-square gives the wrong answer in many instances, including with the data depicted in both Tables 8.1 and 8.2.

To make use of the data in this table to test the hypothesis that the anatomic location of oral lesions is unrelated to the geographical region, we employ the following R commands:

```
> data =read.table("http://statcourse.
  com/research/lesion.csv", sep=",",
  header = TRUE)
> fishers.test(data)
Fisher's Exact Test for Count Data
data: data
p-value = 0.01010
alternative hypothesis: two.sided
```

The final line of the output is meaningless for there are many alternatives, not just two sides, and is due to an error in the `fishers.test` program.

A number of alternate statistics (corresponding to an equal number of alternate hypotheses) are available in the case of contingency tables with more than two rows and columns. Provision for their analysis is not available in R. The interested reader is directed to Good (2006, pp. 119–127) for further guidance.

Binomials

To compare two binomial populations, make use of Fisher's Exact Test for 2 × 2 contingency tables as described in the previous section.

Rare Events (Poisson)

Recently, I had the opportunity to participate in the conduct of a very large-scale clinical study of a new vaccine. I'd not been part of the design team, and when I read over the protocol, I was stunned to learn that the design called for inoculating and examining 100,000 patients, 50,000 with the experimental vaccine, and 50,000 controls with a harmless saline solution!

Why so many? The disease at which the vaccine was aimed was relatively rare. In essence, we would be comparing two *Poisson* distributions. Suppose we could expect 0.8% or 400 of the controls to contract the disease, and 0.7% or 350 of those vaccinated to contract it. Put another way, if the vaccine was effective, we would expect 400/750th of the patients who contracted the disease to be controls. If the vaccine was ineffective (and innocuous), we would expect 50% of the patients who contracted the disease to be controls. Once again, the problem boils down to a single binomial.

We would reject the null hypothesis only if more than qbinom (.95,750,.5) or 398 of those who contracted the disease were controls.

In fact, less than a hundred of those inoculated—treated and control—contracted the disease. The result was a test with extremely low power. As always, the power of a test depends not on the number of subjects with which one starts a trial but the number with which one ends it.

Measurements

To compare the means of two samples, proceed as follows:

1. Decide if your observations are paired or unpaired.
2. Decide whether your alternative to the null hypothesis is one-sided or two-sided.
3. Use the R function t.test() to make the comparison.

To compare teaching methods, 10 school children were first taught by conventional methods, tested, and then taught comparable material by an entirely new approach. The following are the test results:

conventional =c(65,79,90,75,61,85,98,80,97,75)
new = c (90,98,73,79,84,81,98,90,83,88)

Are the two teaching methods equivalent in result? As this is a matched pairs experiment, we need to work with the differences new—conventional. As we will introduce a new method which is sure to entail retraining, new texts, and other costs, only if it represents improvement, a one-sided test is required.

```
> t.test(new,conventional,alternative="g",
    paired=TRUE)
  Paired t-test
data: new and conventional
t = 1.2666, df = 9, p-value = 0.1186
alternative hypothesis: true difference in
 means is greater than 0
95 percent confidence interval:
 -2.639019 Inf
sample estimates: mean of the
 differences 5.9
```

In this second experiment to compare teaching methods, 10 school children were taught by conventional methods, while 10 other children were taught comparable material by an entirely new approach. The following are the test results:

conventional = c(65,79,90,75,61,85,98,80,97,75)
new = c (90,98,73,79,84,81,98,90,83,88)

Are the two teaching methods equivalent in result? This is a nested experiment. As we will introduce a new teaching method, which is sure to entail retraining, new texts, and other costs, only if it represents improvement, a one-sided test is required.

```
> t.test(new,conventional,alternative="g",
   var.equal=TRUE)
  Two Sample t-test
data: new and conventional
t = 1.2672, df = 18, p-value = 0.1106
alternative hypothesis: true difference in
 means is greater than 0
95 percent confidence interval:
 -2.173489 Inf
sample estimates:
mean of x mean of y
  86.4   80.5
```

Additive or Percentage Changes?

The preceding examples assume that the change that results from a new treatment corresponds to a shift in the original distribution, that is, if F is the cumulative distribution before and G is the cumulative distribution after the new treatment, then $G[x] = F[x - \mu]$, where μ is the magnitude of the change. In particular, the expected value of X is increased by μ as are the values of each of the percentiles of the distribution.

But what if the change corresponds to a percentage increase in the expected value? The straightforward solution is to take the logarithm of all values before applying the t-test.

Case-Control Studies

Exceptions to these rules arise in epidemiology when searching for causal factors in case-control studies, for the matching of potentially confounding factors between control and treatment subsets must be done after the data has already been collected. Matching variables:

- Should be chosen before the data is examined.
- Should be limited to variables believed to confound the relationship between exposure and disease. For example, if the matching variable is not associated with disease but is associated with exposure, this will increase the variance of the estimator compared to an unmatched design.
- Should not be strongly correlated with one another.
- Should not be an outcome variable.
- Should not be along the causal pathway between disease and exposure.

By being highly selective and using as few variables as possible, one may satisfy all these restrictions.

Comparing Three or More Populations

The optimal test for comparing multiple treatments depends on whether the treatment categories are unordered (several different fertilizers) or ordered (several different levels of the same fertilizer).

Unordered Treatments

Suppose that crop yield is a function of the fertilizer used plus a random component that is the cumulative result of a large number of factors that we cannot control, such as differences in soil pH, clay concentration, and so forth. In symbols, our *additive* model is that

$$X_{ij} = \mu + f_i + z_{ij}$$

where X_{ij} is the yield of the jth planting treated with the ith fertilizer, f_i is the effect of the ith fertilizer, $\Sigma f_i = 0$, and the experimental errors $\{z_{ij}\}$ are independent of one another and all come from the same probability distribution whose mean is zero. The expected value of X_{ij} is $\mu + f_i$.

Two possible test statistics suggest themselves:

$$F_2 = \sum_{i=1}^{I} n_i \left(\overline{X}_i \right)^2 = \sum_{i=1}^{I} \left(\sum_{j=1}^{n_i} X_{ij} \right)^2 \Big/ n_i$$

$$F_1 = \sum_{i=1}^{I} n_i \left| \overline{X}_i - \overline{X} \right|$$

where $\overline{X}_i$ is the mean of the ith treatment group and n_i the number of observations, and where $\overline{X}$ denotes the mean of all the observations.

F_2 emphasizes the effect of large deviations from the mean, whereas F_1 treats all deviations equally.

Suppose we wish to compare the effects of three brands of fertilizer on crop yield:

FastGro = c(27, 30, 55, 71, 18)
NewGro = c(38, 12, 72)
WunderGro = c(75,76,54)

Note that we have unequal numbers in the various treatment categories.

To analyze this data by means of a permutation test using R, we need two helper functions:

```
F2=function(size,data){
#size is a vector containing the sample
 sizes
#data is a vector containing all the data
 in the same order as the sample sizes
stat=0; start=1
for (i in 1:length(size)){
```

```
      gM = mean(data[seq(from = start,
       length =size[i])])
      stat = stat + size[i]*gM*gM
      start = start + size[i]
      }
   return(stat)
}


F1=function(size,data){
#size is a vector containing the sample
 sizes
#data is a vector containing all the data
 in the same order as the sample sizes
stat=0; start=1
grandMean = mean(data)
for (i in 1:length(size)){
   groupMean = mean(data[seq(from = start,
    length =size[i])])
   stat = stat + abs(groupMean-grandMean)
   start = start + size[i]
   }
 return(stat)
}
```

The main R function we need for the analysis is

```
# One-way analysis of unordered data via a
Monte Carlo
oneway=function(size, data, option=2){
    if (option==1) f1=F1(size,data) else
     f2=F2(size,data)
#Number MC of simulations determines
 precision of p-value
    MC = 1600
    cnt = 0
    for (i in 1:MC){
        pdata = sample (data)
        if (option==1){
            f1p=F1(size,pdata)
            if (f1 <= f1p) cnt=cnt+1
            } else{
```

```
        f2p=F2(size,pdata)
            if (f2 <= f2p) cnt=cnt+1
        }
    }
    #include the original data in the
     count
    cat("p=" ,format((cnt+1)/(MC+1),
     digits=3))
}
```

The decision as to whether to compute F1 or F2 must be made *before* the data is analyzed; otherwise the *p* values are meaningless.

```
> data = c(FastGro, NewGro, WunderGro)
> size =c(length(FastGro), length(NewGro),
  length(WunderGro))
> oneway(size,data,option=1)
p= 0.223
> oneway(size,data,option=2)
p=0.243
```

Ordered Treatments

Pathologists often assess mutagenic potential by observing the number of chromosome abnormalities. The null hypothesis would be that the compound under suspicion is not a mutagen and will not induce chromosome abnormalities, while the ordered alternative would be that the number of abnormalities is an increasing function of the dose.

The statistics F_1 and F_2 offer protection against a broad variety of alternatives, but they do not provide a most powerful test against the ordered alternative that is our specific interest.

In this example, the preferred test statistic is the Pearson correlation, $\sum_{i=1}^{I} f[d_i] \sum_{j=1}^{n_i} y_{ij}$ where d_i is the dose, $f[i] = \log[dose_i + 1]$, and y_{ij} is the number of abnormalities.

TABLE 8.3   Micronuclei in Polychromatophilic
Erythrocytes and Chromosome Alterations
in Bone Marrow of CY-treated Mice Observed
by Frank et al. (1978)

Dose (mg/kg)	Number of animals	Micronucleii per 200 cells	Breaks per 25 cells
0	4	0 0 0 0	0 1 1 2
5	5	1 1 1 4 5	0 1 2 3 5
20	4	0 0 0 4	3 5 7 7
80	5	2 3 5 11 20	6 7 8 9 9

Applying this test to the data in Table 8.3,

```
> dose=c(rep(0,4),rep(5,5),rep(20,4),
  rep(80,5))
> micro=c(0, 0,0, 0, 1,1,1,4,5, 0,0,0,4,
  2,3,5,11,20)
> cor.test(log(dose+1), micro, alternative
  ="g",method="pearson")

  Pearson's product-moment correlation
data: log(dose + 1) and micro
t = 2.5795, df = 16, p-value = 0.01008
alternative hypothesis: true correlation
 is greater than 0
95 percent confidence interval:
 0.1802400 1.0000000
sample estimates: cor 0.5419625
```

Ordinal Data

Often, our data is measured on an ordinal scale such as
"very satisfied," "satisfied," "not unhappy," and so forth.
To assess whether such ratings are meaningful for prediction purposes, a numeric scale needs to be assigned.
Consider the data in Table 8.4: How should we assign a
scale? Are two drinks twice as likely to cause a fetal mal-
formation as one drink?

TABLE 8.4 Maternal Alcohol Consumption (drinks/day)

	0	<1	1 to 2	3 to 5	>6
Malformation absent	117,066	14,464	788	126	37
Malformation present	48	38	5	1	1

Source: Data Gathered by Graubard and Korn (1987).

The possibilities for a scoring method include all of the following:

1. Category number: 0 for the first category, 1 for the second, and so forth
2. Midrank scores. In this example, "no drinks" would be assigned the value (17,066 + 48)/2.
3. Scores determined by the domain expert who collected the data—in this example, a physician or a physiologist

The choice is up to the investigator; but it *must be made before* the data is analyzed if the *p*-values are to be meaningful. Suppose we were to adopt the first of these scoring systems to analyze the data of Table 8.3.

```
> score=c(0,1,2,3,4)
> absent=c(17066,14464,788,126,37)
> present=c(48,3,5,1,1)
> incidence=present/(absent+present)
> cor.test(score, incidence, alternative=
  "g",method="pearson")
 Pearson's product-moment correlation
data: score and incidence
t = 2.6843, df = 3, p-value = 0.03739
alternative hypothesis: true correlation
 is greater than 0
95 percent confidence interval:
 0.05890286 1.00000000
sample estimates: cor 0.8402603
```

Experimental Designs

In this section, we consider the analysis of the more complex designs we looked at in Chapter 4.

Cross-Over Designs

When subjects inevitably drop out of trials so that the design becomes unbalanced, or if treatments are administered over more than two time periods, a permutation test is recommended.

Fourteen children were enrolled in a trial to investigate the effects of two bronchodilators, formoterol and salbutamol, in the treatment of asthma. The outcome variable is peak expiratory flow rate (liters per minute). Alas, one of the children dropped out, leaving the design unbalanced with unequal numbers in the two treatment sequences. Nonetheless, the data can be analyzed via permutation means to produce an exact significance level. Let C represent the result of treating with formoterol, and T the result of treating with salbutamol. Following Good and Xie (2008), our test statistic is ΣC_i. The permutation distribution is obtained by rearranging the *sequence labels* on the subjects so that the number of subjects $\{n_i\}$ in each treatment sequence is preserved.

Begin by downloading the data

```
> con=url("http://statcourse.com/research/
    cross.dat")
> load(con)
```

We suggest you examine the data before beginning to analyze it using the command

```
> edit(cross)
> attach (cross)
```

This last command allows us to address the individual variables in the cross data frame by name.

```
> stat0=0
> for (i in 1:length(subject)) stat0=stat0
  + ifelse(rseq[i]==1,pefrF[i],pefrS[i])
> MC = 1600
> cnt = 0
> for (j in 1:MC){
>        stat=0
>        rseq=sample(seq)
>          for (i in 1:length(subject))
>            stat = stat + ifelse(rseq[i]
              ==1,pefrF[i],pefrS[i])
>          if (stat0 <= stat) cnt=cnt+1
>          }
> #include the original data in the count
> cat("p=" ,format((cnt+1)/(MC+1),
  digits=3))
p= 0.255
```

Complete Multifactor Design

Table 8.5 depicts the effects of two factors, sunlight and fertilizer, on crop yield. The design is balanced, that is, there are three observations for each factor combination.

Let us enter the data from Table 8.5 in R, a treatment combination at a time:

TABLE 8.5   Effect of Sunlight and Fertilizer on Crop Yield

	Fertilizer		
Sunlight	Low	Medium	High
Low	5	15	21
	10	22	29
	8	18	25
High	6	25	55
	9	32	60
	12	40	48

```
sun = c(rep ("L",9),rep ("H",9))
fert = c(rep("L",3),rep ("M",3),rep("H",3),
 rep("L",3),rep ("M",3),rep("H",3))
yield = c(5,10,8,15,22,18,21,29,25,6,9,12,
 25,32,40,55,60,48)
```

To analyze not only the data in this table but the data we'll encounter in the balance of this chapter, in each instance, it is first necessary to specify a model. In this case, the model will be

yield = constant + Effect of Sunlight + Effect of Fertilizer + Interaction of Sunlight and Fertilizer + random residual.

To see why an interaction term is necessary, consider what would happen if we attempted to grow corn in the absence of sunlight or without fertilizer. The corn would not grow in the first case and would be stunted in the second case. Corn requires both sunlight and fertilizer in order to grow. And while sunlight and fertilizer both have a main effect on corn yield, the magnitude of the effect will depend on the level of the other factor. Perhaps I should write "may depend," as in the following analysis we will test to see whether the interaction term is significantly different from zero. To provide for this test in R, I write sun*fert.

```
> fit=lm(yield ~ sun * fert)
> anova(fit)

Analysis of Variance Table
Response: yield
          Df  Sum Sq  Mean Sq  F value  Pr(>F)
sun        1  997.56   997.56  43.902  2.444e-05 ***
fert       2 2952.44  1476.22  64.968  3.652e-07 ***
sun:fert   2  589.78   294.89  12.978  0.0009986 ***
Residuals 12  272.67    22.72
Signif. codes:
0 '***' 0.001 '**' 0.01 '*' 0.05 '.' 0.1 ' ' 1
```

Not unexpectedly, the main effect of sunlight and fertilizer as well as the interaction between them are statistically significant.

Latin Square

Recall that in a Latin square

- There are equal numbers of rows, columns, and treatments.
- Treatments are assigned at random within rows and columns, with each treatment once per row and once per column.

```
fert = c(rep("none",1), rep("cow
manure",1), rep("pig manure",1),
rep("WunderGro",1), rep("NewGro",1))
seed = c("Amsterdam","Seedgirl","Monsanto",
"Burpee","Park","Monsanto","Burpee",
"Amsterdam","Park","Seedgirl","Burpee",
"Monsanto","Park","Seedgirl","Amsterdam",
"Park","Amsterdam","Seedgirl","Monsanto",
"Burpee","Seedgirl","Park","Burpee",
"Amsterdam","Monsanto")
treat = c(rep("treatA",5),
rep("treatB",5), rep("treatC",5),
rep("treatD",5), rep("treatE",5))
yield = c(42,45,41,56,47, 47,54,46,52,
49,55,52,57,49,45, 51,44,47,50,54, 44,
50,48,43,46)
> data = data.frame(treat, fert, seed,
yield)
> #Use of plus signs is essential when
analyzing a Latin Square
> fit = lm(yield ~ fert + treat + seed,
data=data)
> anova (fit)
```

```
Analysis of Variance Table
Response: yield

          Df  Sum Sq  Mean Sq  F value  Pr(>F)
fert       4   17.76    4.440   0.7967  0.549839
treat      4  109.36   27.340   4.9055  0.014105 *
seed       4  286.16   71.540  12.8361  0.000271 ***
Residuals 12   66.88    5.573
---
Signif. codes:
0 '***' 0.001 '**' 0.01 '*' 0.05 '.' 0.1 ' ' 1
```

Fractional Factorial Design

We begin by bringing the crossdes package into memory

```
> library(crossdes)
```

and then download the data obtained from three replications of a half 3^2 design in six blocks

```
> con=url(
+"http://statcourse.com/research/npkf.dat")
> load(con)
```

We suggest you examine the data before beginning to analyze it using the command

```
> edit(npkf)
```

to see the data depicted in Figure 8.1.

```
> npk.aov=aov(yield~block+N*P*K,npkf)
> summary(npk.aov)

          Df  Sum Sq  Mean Sq  F value  Pr(>F)
block      5  343.29   68.659   6.2132  0.004561 **
N          1  189.28  189.282  17.1287  0.001374 **
P          1   10.01   10.008   0.9056  0.360048
K          1  107.54  107.544   9.7320  0.008861 **
N:P        1   31.99   31.994   2.8952  0.114584
N:K        1   51.34   51.337   4.6457  0.052139 .
P:K        1   10.30   10.299   0.9320  0.353403
Residuals 12  132.61   11.051
Signif. codes:
0 '***' 0.001 '**' 0.01 '*' 0.05 '.' 0.1 ' ' 1
```

Data Editor

	block	N	P	K	yield
1	1	0	1	1	49.5
2	1	1	1	0	62.8
3	1	0	0	0	46.8
4	1	1	0	1	57
5	2	1	1	1	59.8
6	2	1	1	1	58.5
7	2	0	0	1	55.5
8	2	0	1	0	56
9	3	0	1	0	62.8
10	3	1	1	1	55.8
11	3	1	0	0	69.5
12	3	0	0	1	55
13	4	1	0	0	62
14	4	1	1	1	48.8
15	4	0	0	1	45.5
16	4	0	1	0	44.2
17	5	1	1	0	52
18	5	0	0	0	51.5
19	5	1	0	1	49.8
20	5	0	1	1	48.8
21	6	1	0	1	57.2
22	6	1	1	0	59
23	6	0	1	1	53.2
24	6	0	0	0	56

FIGURE 8.1 The effect of nitrogen (N), phosphate (P), and potassium (K) on the yield of pea plants.

Each half design is repeated the same number of times (3), and there are equal numbers of treatment combinations for each plot. Missing observations would have unbalanced the design, thus compromising the entire analysis.

To Learn More

The details of Student's t-test may be found in virtually every statistics textbook published in the past century. The binomial test for comparing two Poisson distributions is due to Lehmann (1986). For more on the inverse binomial, see Hilbe (2011).

The value of the permutation test in analyzing the one-way layout was established by Good and Lunneborg (2006). The distribution-free nature of Pearson's correlation is demonstrated by Good (2009).

Case-control study guidelines will be found in Beslow and Day (1980), Schlesselman (1982), and Vandenbroucke et al. (2007).

Jagers (1980) shows that for balanced designs the analysis of variance F statistic is almost distribution-free.

CHAPTER 9

Multiple Variables
and Multiple Tests

Prescription

- When you've made multiple simultaneous observations on several variables, consider employing a multivariate test such as Hotelling's T^2.
- When you perform a series of tests, it's essential that you control not merely the individual but the overall probability of making Type I errors.

Multiple Variables

Basing an analysis on simultaneous observations on several variables, such as height, weight, blood pressure, and cholesterol level, enables us to detect subtle changes that might otherwise not be detectable except with very large, prohibitively expensive samples.

We may use Hotelling's T^2 for two-sample comparisons of measurements in the multivariate case. This statistic weighs the contribution of individual variables and pairs of variables in inverse proportion to their covariances. This has the effect of rescaling each variable so that the most weight is given to those variables that can be

109

measured with the greatest precision, and those variables that can provide information not provided by the others.

Unlike Student's-*t*, Hotelling's T^2 is highly sensitive to departures from normality, and a distribution-free permutation approach to determining a *p*-value is recommended. To perform a permutation test with multivariate observations, we treat each vector of observations on an individual subject as a single indivisible entity. When we exchange treatment labels, we relabel on a subject-by-subject basis so that all observations on a single subject receive the same new treatment label.

Proceed in three steps:

1. Compute Hotelling's T^2 for the original observations.
2. Compute Hotelling's T^2 for each relabeling of the subjects.
3. Determine the percentage of relabelings that lead to values of the test statistic that are as or more extreme than the original value.

```
#due to Nick Horton
hotelling = function(y1, y2) {
  # check the appropriate dimension
  k = ncol(y1)
  k2 = ncol(y2)
  if (k2!=k)
    stop("input matrices must have the
     same number of columns: y1 has ",
     k, " while y2 has ", k2)

  # calculate sample size and observed
  means
  n1 = nrow(y1)
  n2 = nrow(y2)
  ybar1= apply(y1, 2, mean);
   ybar2= apply(y2, 2, mean)
  diffbar = ybar1-ybar2

  # calculate the variance of the
   difference in means
```

```
v = ((n1-1)*var(y1)+ (n2-1)*var(y2)) /
  (n1+n2-2)

# calculate the test statistic and
  associated quantities
t2 = n1*n2*diffbar%*%solve(v)%*%diffbar/
  (n1+n2)
f = (n1+n2-k-1)*t2/((n1+n2-2)*k)
pvalue = 1-pf(f, k, n1+n2-k-1)
# return the list of results
return(list(pvalue=pvalue, f=f, t2=t2,
  diff=diffbar))
}
#Compute the exact p-value
photell= function
  (samp1,samp2,fir,last,MC){
#fir=first of consecutive variables to be
 included in the analysis
#last=last of consecutive variables to be
 included in the analysis
#compute lengths of samples
d1=dim(samp1)[1]
d2=dim(samp2)[1]
d=d1+d2
samp=c(samp1,samp2)
#create dataframe to hold rearranged data
nsamp=samp
#compute test statistic for data
res=hotelling(samp[1:d1, fir:last],
 samp[(1+d1):d,fir:last])
stat0=as.numeric(res[[3]])
#run Monte Carlo simulation of p-value
cnt=1
for (j in 1:MC){
  nu=sample(lgth)
  for (k in 1:d)nsamp[nu[k],]=samp[k,]
  res=hotelling(nsamp[1:d1, fir:lst],nsamp
  [(1+d1):d, fir:last])
  if (as.numeric(res[[3]])>=stat0)
   cnt=cnt+1
  }
```

```
return (cnt)
}
print(c("pvalue=",cnt/MC), quote=FALSE)
```

Repeated Measures

In many experiments, we study the development of a process over a period of time, such as the growth of a tumor or the gradual progress of a cure. If our observations are made by sacrificing different groups of animals at different periods of time, then time would simply become another variable in the analysis which we may treat as a covariate. But if all observations are made on the same subjects, then the multiple observations on a single individual will be interdependent. And all the observations on a single subject must be treated as a single multivariate vector.

Although normality of these observations might be assumed, the variances and covariances surely vary with time so that the two-way ANOVA model is not appropriate. Standardize the observations, subtracting the baseline value at $t = 0$ from each one, so that at each point in time, the resulting differences are exchangeable. Treat the transformed observations on each subject as a multivariate vector $(X[t_1] - X[t_0], X[t_2] - X[t_1], . . .)$, so that we can compare the two treatments using Hotelling's T^2.

Here is an example: Higgins and Noble (1993) analyze an experiment whose goal was to compare two methods of treating beef carcasses in terms of their effect on pH measurements of the carcasses taken over time. Treatment level B is suspected to induce a faster decay of pH values.

Observed data are shown in Table 9.1.

Suppose we read this table into R's memory as a spreadsheet

```
> data= read.table ("c:/AnalData/repMes.
  csv",sep=",", header=T)
> attach(data)
```

TABLE 9.1   pH as a Function of Time

Group	t0	t1	t2	t3	t4	t5
A	6.81	6.16	5.92	5.86	5.80	5.39
A	6.68	6.30	6.12	5.71	6.09	5.28
A	6.34	6.22	5.90	5.38	5.20	5.46
A	6.68	6.24	5.83	5.49	5.37	5.43
A	6.79	6.28	6.23	5.85	5.56	5.38
A	6.85	5.51	5.95	6.06	6.31	5.39
B	6.64	5.91	5.59	5.41	5.24	5.23
B	6.57	5.89	5.32	5.41	5.32	5.30
B	6.84	6.01	5.34	5.31	5.38	5.45
B	6.71	5.60	5.29	5.37	5.26	5.41
B	6.58	5.63	5.38	5.44	5.17	5.62
B	6.68	6.04	5.62	5.31	5.41	5.44

```
> nudata=data.frame(t1-t0,t2-t0,t0-t3,
  t0-t4,t0-t5)
> photell(nudata,fir=1,last=5,MC=1600)
```

Multiple Tests

With large-scale studies such as clinical trials as well as in the analysis of biomedical images, so many variables are under investigation that one or more of them are practically guaranteed to be significant by chance alone. If we perform 20 tests at the 5% or 1/20 level, we expect at least one (potentially false) significant result on average. If the variables are related (and in most large-scale medical and sociological studies, the variables have complex interdependencies), the number of falsely significant results could be many times greater. In this section, you'll learn methods for controlling either the family-wise error rate or the false discovery rate. The latter provides greater power than does controlling the family-wise error rate at a cost of increasing the likelihood of obtaining Type I errors.

Controlling the Family-Wise Error Rate

A resampling procedure outlined by Troendle (1995) allows us to work around the dependencies among tests. Suppose we have measured k variables on each subject, and are now confronted with k test statistics using one of the methods in the preceding chapter. To make these statistics comparable, we need to standardize them and render them dimensionless, dividing each by its standard error. For example, if one variable, measured in centimeters, takes values such as 144, 150, and 156, and the other, measured in meters, takes values such as 1.44, 1.50, and 1.56, we might divide each of the first set of observations by 4, and each of the second set by 0.04.

Next, we order the standardized statistics by magnitude, that is, from smallest to largest. We also reorder and renumber the corresponding hypotheses. The probability that at least one of these statistics will be significant by chance alone at the 5% level is $1 - [1 - 0.05]^k$. But once we have rejected one hypothesis (assuming it indeed was false), there will only be $k - 1$ true hypotheses to guard against rejecting.

1. Focus initially on the largest of the k test statistics, and repeatedly resample the data, with or without replacement, to determine the p-value.
2. If this p-value is less than the predetermined significance level, then accept this hypothesis as well as all the remaining hypotheses.
3. Otherwise, reject the corresponding hypothesis, remove it from further consideration, and repeat steps 1 to 3.

Controlling the False Discovery Rate

Suppose we wish to control the expected proportion of incorrectly rejected null hypotheses or type I errors at $q < 1$. Begin by sorting the p-values corresponding to the m univariate hypotheses as $p[1] \leq \ldots \leq p[m]$. Set the cutoff values at $c_i = iq/\{m + 1 - i[1 - q]\}$, $i = 1, \ldots, m$.

Let $k = \max \{1 \le i \le m$ with $p[j] \le c_i, j = 1, \ldots, i\}$. Reject the k hypotheses associated with $p[1] \ldots p[k]$ if k exists; otherwise, accept all hypotheses.

```
CFDR = function(p, q) {
m = length(p);
p = sort(p);
i = 1;
while(i<=m && p[i] <= i*q/(m+1-i*(1-q)))
 i = i+1;
i-1
}
```

We've not specified which of the univariate tests in Chapters 8 or 10 were employed to obtain the individual *p*-values, as this will depend upon your specific application.

To Learn More

Good (2011; Chapter 5) discusses several alternative multivariate tests. Good (2011; Chapter 8) provides additional methods for controlling the error rate.

CHAPTER 10

Miscellaneous Hypothesis Tests

Prescription

The hypothesis tests in this chapter have this in common: the observations are not exchangeable. In most instances, the decision to reject or accept is based on a confidence interval. In others, the only solution is to employ a bootstrap.

Hypothesis Tests and Confidence Intervals

Confidence intervals can be derived from the rejection regions of our hypothesis tests, whether the latter are based on parametric or nonparametric methods. Vice versa, hypothesis tests can be derived from confidence intervals.

Suppose $A(\theta')$ is a $1 - \alpha$ level acceptance region for testing the hypothesis $\theta = \theta'$; that is, we accept the hypothesis if our test statistic T belongs to the acceptance region $A(\theta')$ and reject it otherwise. Let $S(X)$ consist of all the parameter values θ^* for which $T[X]$ belongs to the acceptance region $A(\theta^*)$. Then $S(X)$ is an $1 - \alpha$ level confidence interval for θ based on the set of observations $X = \{x_1, x_2, \ldots, x_n\}$.

The probability that *S(X)* includes θ_o when $\theta = \theta_o$ is equal to $\Pr\{T[X] \; \varepsilon \; A(\theta_o) \text{ when } \theta = \theta_o\} = 1 - \alpha$.

As our confidence $1 - \alpha$ increases, the width of the resulting confidence interval increases. Thus, a 95% confidence interval is wider than a 90% confidence interval.

By the same process, the rejection regions of our hypothesis tests can be derived from confidence intervals. Suppose our hypothesis is that the odds ratio for a 2×2 contingency table is 1. Then we would accept this null hypothesis if and only if our confidence interval for the odds ratio includes the value 1.

We apply this same technique in the next section to a test of equivalence.

Testing for Equivalence

Suppose we hope to bring a generic version of a drug to market. (Consumers hope we'll be successful, also, as they'll be able to pay less.) The FDA insists that we first demonstrate that our version is "equivalent" to the old, that is, that its use produces "approximately" the same effects.

Two observables are said to be "equivalent" if their expected values lie within a prespecified amount of one another, an amount the FDA's panel of experts would agree is equivalent in practice.

Surprisingly, the standard *t*-test won't do to test for equivalence. We can't assume two formulations or two processes are "equivalent" merely because we fail to reject the null hypothesis. And, if we make the samples large enough, we will always reject the null hypothesis.

What is required is a test that rejects if $|EX - EY| >$ epsilon and accepts otherwise.

Let's suppose we have two sets of data, one resulting from the existing or conventional process, the other from a newer, less expensive process.

conventional = c(65,79,90,75,61,85,98,80,97,75)
new = c (90,98,73,79,84,81,98,90,83,88)

To test using R that the expected value resulting from employing the new package lies within epsilon = 3 of the conventional method's expectation, first download and install the equivalence package.

```
> library(equivalence)
> tost(new, conventional, epsilon=3, paired=F)

$mean.diff            [1] 5.9
$se.diff              [1] 4.655821
$alpha                [1] 0.1
$ci.diff              [1] -0.3334087 12.1334087
attr(,"conf.level")   [1] 0.8
$df                   15.44158
$epsilon              [1] 3
$result               [1] "not rejected"
$p.value              [1] 0.7287718
```

The confidence interval for the difference ($ci.diff) contains epsilon, and we conclude that the two methods are equivalent.

When Variables Are Not Identically Distributed

Often, an increase in expected value is accompanied by an increase in variability. But the Student's *t*-test requires that the variables being compared have the same variance. One solution is a variance-stabilizing transformation. For example, when the increase is a constant percentage of the original, taking the logarithms of the observations permits a comparison using Student's-*t*.

Sometimes not even the observations within the same sample have the same variance, for example, when the observations were made by different observers using different apparatuses. Reducing the observations in both samples to a common set of ranks provides a partial solution.

A second solution is to proceed as follows:

1. Take a series of pairs of bootstrap samples, one member of each pair from the first sample and one from the second.

2. Compute the *t*-statistic for each such pair.
3. Repeats steps 1 and 2 to form a confidence interval for the test statistic.
4. Reject the null hypothesis if the confidence interval does not contain zero.

```
#set number of bootstrap samples
N =1600
stat = numeric(N) #create a vector in
 which to store the results
 #the elements of the vector will be
 numbered from 1 to N
#Set up a loop to generate a series of
 bootstrap samples
for (i in 1:N){
  #bootstrap sample counterparts to
  observed samples are denoted with "B"
  OneB= sample (Samp1, replace=TRUE)
  TwoB= sample (Samp2, replace=TRUE)
  res=t.test(OneB,TwoB)
  stat[i] = res[[1]]
  }
quantile(stat,c(.05,.95))
```

Comparing Variances

Precision is essential in a manufacturing process. Items that are too far out of tolerance must be discarded. An entire production line is brought to a halt if too many items exceed (or fall below) designated specifications. With some testing equipment, such as that used in hospitals, precision can be more important than accuracy. Accuracy can always be achieved through the use of standards with known values, while a lack of precision may render an entire sequence of tests invalid.

To compare variances, we execute the following steps:

1. Compute the median for each sample.
2. Replace each of the observations by the square of their deviations about their sample median.

3. Discard the redundant linearly dependent value from each sample.
4. As test statistic, use the sum S of the remaining deviations in the first sample.
5. Derive the permutation distribution of S.

```
Using R
ptest=function(samp1,samp2,MC){
  sizeS=length(samp1)
  first=resid(samp1)
  res=c(first,resid(samp2))
    sumorig=sum(res[1:sizeS])
    cnt=0
  for (i in 1:MC){
    D=sample(res,sizeS)
    if (sum(D)<= sumorig)cnt<-cnt+1
  }

#eliminate smallest (and redundant)
 residual resid=function(data){
  mn=median(data)
  res=abs(data-mn)
  res=sort(res)
  return (res[-1])
  }
```

Cluster Sampling

To reduce the cost of sampling, data is often gathered in clusters from families, fraternities, army units, and other groups who share common values and work or leisure habits.

Unless stratification is appropriate, we must treat each cluster as if it were a single observation, replacing individual values with a summary statistic such as an arithmetic average. Cluster-by-cluster means are unlikely to be identically distributed, having variances, for example, that will depend on the number of individuals that make up the cluster. If there are a sufficiently large number of clusters in each treatment group, draw bootstrap samples composed of one observation drawn at random from each cluster to test for treatment differences.

Group Randomized Trials

Group randomized trials (GRTs) in public health research typically use a small number of randomized groups with a relatively large number of participants per group; examples include work sites, schools, clinics, neighborhoods, even entire towns or states. A group can be assigned to either the intervention or control arm but not both; thus, the group is nested within the treatment.

Not unexpectedly, outcomes within a group are positively correlated. The sampling variance for the average responses in a group is increased by a factor of $(n - 1)\sigma$. The good news is that σ is relatively small as the groups tend to be more homogeneous than the population at large; the bad news is that the inflation factor is a function of the group size.

Most parametric tests yield misleading p-values unless the observations are normally distributed, according to Feng et al. (2001). Gail et al. (1996) have shown that the permutation test remains valid (exact or near exact in nominal levels) under almost all practical situations, including unbalanced group sizes, as long as the number of groups are equal between treatment arms or equal within each block if blocking is used.

Testing for Trend

Suppose we have in hand a series of observations—stock market prices for example, and we wish to test for trend. The null hypothesis is that the magnitude and direction of future changes does not depend upon the current value. If the intervals, $t_i - t_{i-1}$ between observations are meaningful, the optimal test statistic is the Pearson correlation $\mathrm{sum}(t_i X_i)/\mathrm{sd}(t)^*\mathrm{sd}(x)$ described under "Ordered Treatments" in Chapter 8. If they are not, the optimal test statistic is $\sum \left(X_t - X_{t-1} \right)^2$, and the p-value needs to be determined by permutation means.

CHAPTER **11**

Sample Size Determination

Prescription

- Prepare a budget.
- Specify all of the following:
  - Desired power and significance level
  - Smallest practical effect of interest
  - Type of data to be analyzed
  - Whether the sample size is to be determined sequentially
- Compute the desired final sample size.
- Estimate proportion of sample left after recruitment failures, elimination of ineligibles, withdrawals, dropouts, no-shows, and nonresponders.
- Specify starting sample size.

Prepare a Budget

The optimal sample size is determined by the following factors:

- The cost of preparing to gather a sample
- The cost of collecting a sample which depends on the sample size
- The costs associated with making a Type I error

- The cost associated with making a Type II error
- The smallest effect of practical interest

Trade-offs exist among all these factors which make it essential for you to have a firm grasp on both costs and available funds before you begin.

The greater the accuracy and precision of your data, the fewer samples you'll require. But more precise measurements cost more per sample.

To ensure the quality of your data, you need to spend more up front on questionnaire design, training, and pilot trials, and more during the study on monitoring.

The more Type I and Type II errors you are willing to permit, the smaller the necessary sample size. The smaller the smallest effect of practical interest, the larger the sample size it will take to detect it.

Sequential Sampling

Determining sample size as we go (sequential sampling), rather than making use of a predetermined sample size, can have two major advantages:

1. Fewer samples (on the average)
2. Earlier decisions

When experiments are destructive in nature (as in testing condoms) or may have an adverse effect on the experimental subject, it is best not to delay decisions until some fixed sample size has been reached.

Figure 11.1 depicts a sequential trial of a new vaccine after eight patients, who had received either a vaccine or an innocuous saline solution, developed the disease. Each time a control patient came down with the disease, the jagged line was extended to the right. Each time a patient who had received the experimental vaccine came down with the disease, the jagged line was extended upward one notch. This experiment will continue until either of the following occurs:

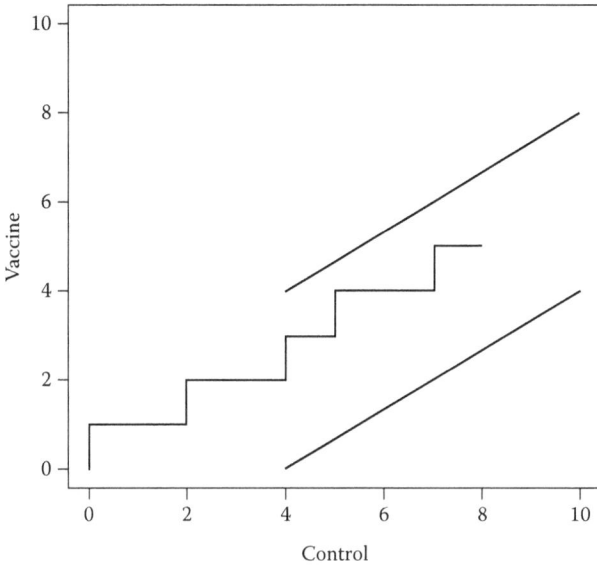

FIGURE 11.1 Sequential trial in progress. (Reprinted from Good, P.I., *Introduction to Statistics via Resampling Methods and R/SPlus,* John Wiley & Sons, Inc., Hoboken, NJ, 2005. With the permission of John Wiley & Sons.)

A. The jagged line crosses the lower boundary—in which case, we will stop the experiment, reject the null hypothesis, and immediately put the vaccine into production

B. The jagged line crosses the upper boundary—in which case, we will stop the experiment, conclude that the vaccine is of negligible value, and abandon further work with it

What Abraham Wald (1950) showed in his pioneering research was that on the average the resulting sequential experiment *would require many fewer observations* whether or not the vaccine was effective than would a comparable experiment of fixed sample size.

If the treatment is detrimental to the patient, we are likely to hit one of the lower boundaries early. If the

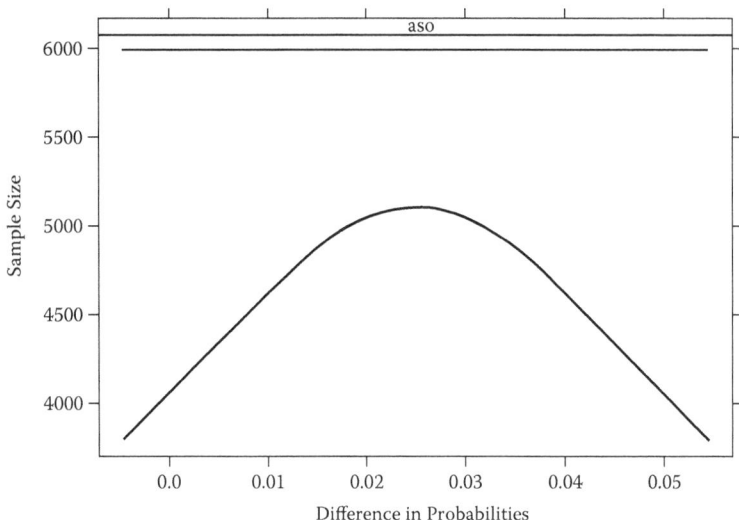

FIGURE 11.2 Average sample size as a function of the difference in probability. (Reprinted from Good, P.I., *Introduction to Statistics via Resampling Methods and R/SPlus,* John Wiley & Sons, Inc., Hoboken, NJ, 2005. With the permission of John Wiley & Sons.)

treatment is far more efficacious than the control, we are likely to hit an upper boundary early. Even if the true difference is right in the middle between our two hypotheses, for example, because the treatment is only 2.5% better when the alternative hypothesis is that it is 5% better, we may stop early on occasion. Figure 11.2 shows the average sample size as a function of the difference in the probabilities of success for each treatment. When this difference is less than 0% or greater than 5%, we'll need about 4000 observations on average before stopping. Even when the true difference is right in the middle, we will stop after about 5000 observations, on average. In contrast, a fixed-sample design requires nearly 6000 observations for the same Type I error and power.

Warning: Simply performing a standard statistical test after each new observation as if the sample size were fixed will lead to inflated values of Type I error. The boundaries depicted in Figure 11.1 were obtained using formulas specific to sequential design. Not surprisingly, these formulas require us to know every one of the same factors we need to know when the experiment is of fixed size.

Final Sample Size
Binomial Trials—Estimating a Proportion

In many instances, there are only two possible outcomes for each observation—success or failure, a vote for our candidate or a vote against. Suppose we desire to cut down a stand of 500-year-old redwoods in order to build and sell an expensive line of patio furniture. Unfortunately, the stand is located on State property. Fortunately, we know a politician we feel could be persuaded to facilitate our purchase of the land. In fact, his agent has provided us with a rate card.

The politician is up for re-election and believes that a series of TV and radio advertisements purchased with our money could guarantee him a victory. He claims that those advertisements would guarantee him 55% of the vote. Our own advisors say he'd be lucky to get 40% of the vote without our ads and the best the TV exposure could do is give him another 5%.

We decide to take a poll. If it looks like only 40% of the voters favor the candidate, we won't give him a dime. If 46% or more of the voters already favor him, we'll pay to saturate the airwaves with his promises. We decide we can risk making a Type I error and back him with probability 5%, and a Type II error and ignore him with probability 10%. That is, if p, the probability that a voter selected at random will favor this candidate is 0.40, then the probability based on our poll of rejecting the hypothesis that $p = 0.40$ should be no greater than 5%. And if $p = 0.46$, then the probability based on our poll of rejecting the hypothesis that $p = 0.40$ should be at least 90%.

To determine how large should our sample must be to ensure these limitations are met, proceed in an iterative fashion:

1. Guess a starting value, say a sample size of 10.
2. Estimate a cutoff qbinom(.95,10,.4) = 7
3. Check the Type II error; pbinom(7,10,.46) = 0.9682895; which is far too large.

Repeat these steps, using a larger sample size, 20, say. qbinom(.95,20,.4) = 12 and pbinom(12,20,.46) = 0.9306301. Not much improvement.

```
qbinom(.95,200,.4)= 91;
   pbinom(91,200,.46) = 0.4724862
qbinom(.95,500,.4) = 218;
   pbinom(218,500,.46) = 0.1510386
qbinom(.95,600,.4) = 260;
   pbinom(260,600,.46) = 0.1019674
```

Perhaps we can get a slight improvement via a small increase in sample size: qbinom(.95,602,.4) = 261 and pbinom(261,600,.46) = 0.1173681. When we increased the sample size, the probability of making a Type II error increased! What's going on?

The explanation lies in the discrete nature of the binomial distribution. As can be seen from Figure 11.3, while the overall trend is an increase in power with increasing sample size, the increase is far from smooth.

Poisson Counts of Rare Events

As we saw in Chapter 8, the power of a test for a Poisson distribution is conditional on the number of events of interest, and cannot be guaranteed in advance. Recently, I had the opportunity to participate in the conduct of a very large-scale clinical study of a new vaccine against pneumonia. I'd not been part of the design team, and was

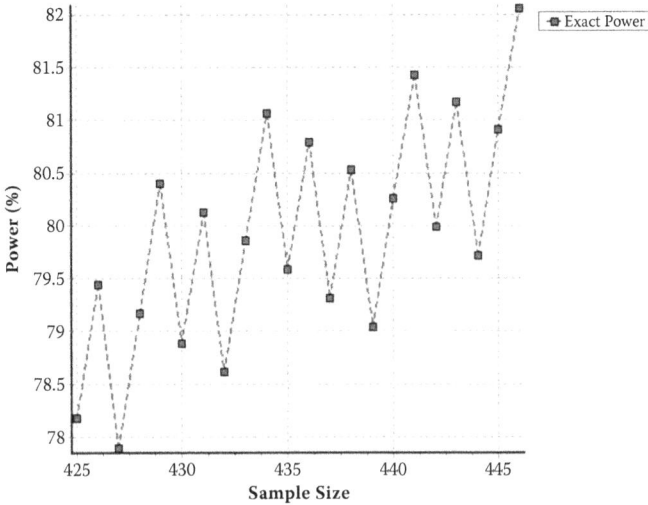

FIGURE 11.3 Estimated power for comparing $p = 0.40$ with $p = 0.46$ at the 5% significance level.

stunned to learn that the design called for inoculating and examining 100,000 service personnel, 50,000 with the experimental vaccine, and 50,000 controls with a harmless saline solution.

The initial calculations suggested one could expect 0.8% or 400 of the controls would contract the disease, and 0.7% or 350 of those vaccinated would contract it. Put another way, if the vaccine was effective, we would expect 400/750th of the patients who contracted the disease to be controls, while if the vaccine was ineffective (and innocuous) we would expect 50% of the patients who contracted the disease to be controls. The problem boils down to a single binomial. We would reject the null hypothesis only if more than qbinom(.95,750,.5) or 398 of those who contracted the disease were controls. The probability of detecting the effect of the vaccine was 1-pbinom(398,750,400/750), or less than 55%.

But those numbers were merely expected or average values; they could not be guaranteed. In fact, pneumonia was relatively rare that year, and less than a hundred of those inoculated—treated and control—contracted the disease.

Measurements

Almost Normal Data

As with all sample size calculations, the number of observations per group will depend on delta, the true difference in means, the standard deviation, the significance level or Type I error probability), the power of the test (1 minus Type II error probability), and the nature of the alternative (one- or two-sided).

In this example, the effect size delta is given in standard deviation units.

```
> power.t.test(delta=1,power=.8)
Two-sample t-test power calculation
  n = 16.71477
  delta = 1
  sd = 1
  sig.level = 0.05
  power = 0.8
alternative = two.sided
Note: n is the number in *each* group
```

Since the sample size must be an integer, let's specify that *n* = 17.

```
> power.t.test(,n=17,delta=1)
Two-sample t-test power calculation
  n = 17
  delta = 1
  sd = 1
  sig.level = 0.05
  power = 0.8070359
alternative = two.sided
NOTE: n is the number in *each* group
```

We might also have specified that type = "paired"
and/or that alternative = "one.sided" as in

```
> power.t.test (n=10, power=0.8, type=
  "paired", alternative="one.sided")
Paired t test power calculation
  n = 10
  delta = 0.8528138
  sd = 1
  sig.level = 0.05
  power = 0.8
alternative = one.sided
NOTE: n is the number of *pairs*, sd is
std.dev. of *differences* within pairs
```

Strictly Empirical Distributions

When we aren't sure about the underlying distribution or,
as in the example that follows, are absolutely sure that the
distribution from which the observations are drawn is not
in the least like a bell-shaped or survival curve, we can
still *simulate* an answer using bootstrap samples.

A client of mine was in the business of refurbishing
certain medical devices. They needed to convince the
FDA that their refurbished devices were every bit as good
as the original. Specifically, they needed to convince the
FDA that test results with their refurbished devices were
equivalent to test results with original equipment. My client had plenty of refurbished devices on hand, but they
were highly reluctant to spend the big bucks to purchase
the necessary never-used models.

The results for the few refurbished units that had been
tested—the tests were destructive—were best depicted
as a U-shaped distribution taking values on a six-point
scale, the very opposite of the continuous bell-shaped
normal. I asked what the distribution of a defective sample of such units might look like and was shown a similar
picture, but with the majority of units shifted from the
right- or high-end to the left- or low-end of the U.

The FDA would not be satisfied unless the sampling scheme we came up with would reject samples from the latter distribution at least 80% of the time. So much for the samples of size one (1) that my client would have preferred to see. On the other hand, the samples of size nine (9) called for by the use of a normal approximation seemed unnecessarily large, especially since this meant the destruction of nine units of each type of device that was to be refurbished!

A computer was used to simulate the taking of samples of various sizes from the existing set of test results as well as from the worst-case distribution. We proceeded in iterative fashion:

1. Set the sample size.
2. Draw repeated bootstrap samples from the existing test results to establish a cutoff value for rejection at the 5% level.
3. Simulate the drawing of repeated samples from the worst-case distribution to establish the power associated with that cutoff value.

Table 11.1 illustrates our findings using 50 to 400 bootstrap samples.

We took 50 bootstrap samples when just trying to get a rough idea of the correct sample size and 400 when closing in on the final value. When sampling from the worst-case distribution, the number of rejections divided by the number of simulations provides an estimate of the power for the specific experimental design and sample size. As always, a larger sample size meant greater power.

TABLE 11.1 Simulation Results ($\alpha = 5\%$)

Sample Size	3	4	5	6
Cutoff for Mean	5.5	5.2	5	4.8
Power (%)	65	72	80	85

Analyzing the One-Way Layout

Using R, we can easily obtain an estimate of the necessary sample sizes for a balanced one-way analysis of variance test, even if later on missing observations force us to use a permutation test to do the actual analysis. We need to specify all of the following: number of groups, smallest effect size of interest (squared), within-group variance, significance level, and the power of the test.

The within-group variance will usually be taken as the variance of any of the treatment groups for which we already have some pilot data.

```
> power.anova.test(groups = 4, between.var
 = 1, within.var = 1,sig.level = 0.05,
 power = 0.8)
Balanced one-way analysis of variance
 power calculation groups = 4
n = 4.734463
between.var = 1
within.var = 1
sig.level = 0.05
power = 0.8
NOTE: n is the number in each group
```

Initial Sample Size

The initial sample size will need to be a multiple of the required final sample size in order to correct for the possibility of missing data.

Let's assume you've practiced all the preventive measures described in Chapter 5. Even so, on the day the study is to commence, you realize that you have only half the necessary amount of reagent and that one of your monkeys is running a fever. A physician who'd promised to deliver a half dozen patients quits to join Doctors Without Borders, and his partner comes up with only a fraction of the patients you hoped he would see. You send out a hundred questionnaires, but only 75 are returned and of these, 25 or so are incomplete. To attain the desired power for your tests (or, equivalently, the desired precision for your estimates) you are going to have to budget for a much larger initial sample in order to attain the desired final number at the conclusion of your study.

Animal Experiments

If unable to start an animal experiment with the desired number of subjects, whether because of a shortage of subjects or of materials, the solution is to perform the experiment in several stages, allowing for at least one additional animal per stage over and above the number originally projected.

Only experience will let you know whether to include one or two extra animals or cages to account for losses during the trials. (Yes, cage tops do drop on hamsters' heads.)

Mean-Time-to-Failure Trials

If you can't start with the desired number of units, then consider doing a sequential experiment.

Clinical Trials

Clinical trials are plagued by recruitment failures, ineligibles, withdrawals, dropouts, no-shows, and nonresponders. A comprehensive list of preventive measures is provided in Good (2005).

Surveys

The chief sources of reduced sample size in surveys are the following:

> *Ineligible subjects.* Forms were sent to or, despite your careful instructions, interviews were conducted with individuals who are not part of your study population.
> *Dropouts* plague longitudinal studies. Retention rates as low as 40% after one year have been reported.

Nonresponders. Study after study has shown that individuals who do not return forms or leave certain questions unanswered may have quite different attitudes from those who do. The only solution is to subsample the nonresponders to document the differences. (Note that the cost per nonresponder may be far in excess of the original cost you projected due to the need for personal contact.)

To Learn More

For the sample size for clinical trials, see Machin et al. (2008). To construct bounds for group sequential trials, download and install the `Hmisc` R library, then make use of `ldBands()`.

For engineering studies, see Mathews (2010). On sources and frequency of missing data, see Lang and Secic (1997).

Part IV

BUILDING A MODEL

Prescription

The reader who has already completed a course in statistics may feel that he or she already knows all about regression Just apply least squares to obtain the best-fitting line between the dependent and the independent variables. But not all relationships you will model are linear, nor will you always want to estimate the mean. And least squares is not always the best method; model building can be far more complex.

Fortunately, you won't need to memorize and apply a bunch of complex formulas to build your models. The chapters in this section will guide you to the appropriate techniques along with the software you'll need to apply them.

Before you can construct a model, you need to answer a few basic questions:

1. What variables do you wish to predict? Are they counts, measurements, or times to events?
2. What is the nature of the relationship between the outcome to be predicted and the predictors?
3. Will you be able to measure the predictors without error?

4. How are the residual errors distributed?
5. Over what ranges do you wish to make your predictions?
6. What are the principal predictors?

Which Variables Do You Wish to Predict?

Consider predicting *effects* that are in a cause-and-effect relationship with the predictors. Some statistics texts refer to them as the *dependent variable,* others as the outcome or *outcome variable.* If you're not certain which variable is of which type, then you need to make an immediate decision: The model obtained by assuming that $EY = f[X]$ may be quite different from the model one obtains assuming $Ef[X] = Y$, where $f[\,]$ is a known function, for example, X^2, or $\log[X]$, or just X; see Figure IV.1.

By EX, we mean the expected value of X. If a variable X can take a finite number of possible values, then $EX = \sum kp(X = k)$, where $p(X = k)$ denotes the probability that $X = k$.

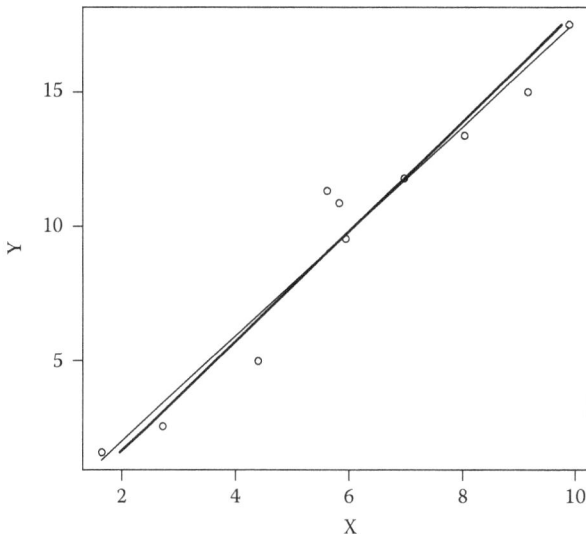

FIGURE IV.1 Comparison of the models $EY = f(X)$ and $Ef[X] = Y$.

Nature of the Relationship

Is the relationship between cause and effect explicit as in $EY = aX + b$? Or require a link function as in $L[EY] = aX + b$, where L is a specified function of the variables X and Y? If the latter, see the section in Chapter 13 entitled "Generalized Linear Models."

Do you wish to predict the expected value or the median value and other quantiles of the outcome variable's distribution? If the latter, see Chapter 12 in the section entitled "Quantile Regression."

If the predictors are likely to be measured with significant errors, see Chapter 12 in the section entitled "Errors-in-Variables Regression."

If the predicted value is a count, see Chapter 13 in the sections entitled "Poisson Regression" and "Binomial Regression."

If the predicted value is a probability or a frequency, see Chapter 13 in the section entitled "Logistic Regression."

If the value to be predicted is the time to an event, see Chapter 13 in the section entitled "Modeling Survival Data."

If the predicted value is continuous (though not a probability or the time to an event) and the relationship takes the explicit form $EY = AX$, consult our discussion in the next chapter.

When your objective is classification rather than regression, or your predictors are interdependent and their interaction may lead to reinforcing synergistic effects, or a mixture of continuous and categorical variables, highly skewed data, and large numbers of missing observations adds to the complexity of the analysis, consider using decision trees as described in Chapter 14.

CHAPTER **12**

Ordinary Least Squares

Prescription

In this chapter, we shall be concerned with predicting variables that can take a continuous possible range of values. Measurements such as weight, height, crop yield, and survival time are all examples. Our models take the explicit form $EY = AX$, where Y denotes the outcome, that is, the effect, X is a matrix of the predictors, the causes, and A is a vector of the coefficients to be estimated.

You'll learn how to use R to fit a linear regression line with one or many predictors and to provide confidence intervals for model coefficients. You'll learn five different ways of improving the fit of the model to the data, and you will also learn to use the bootstrap to validate the results.

Linear Regression

The class of linear regression models includes *all* of the following:

$$T_{ij} = \mu + \beta\, t_i + z_{ij}$$
$$T_{ij} = \mu + \beta\, t_i + \gamma\, t_i^2 + z_{ij}$$
$$T_{ij} = \mu + \beta\cos\,(2\pi^*(t_i + 300)/1440) + z_{ij}$$

where T_{ij}, the dependent variable, represents the body temperature of an individual j at the t_ith minute after midnight, the error terms $\{z_{ij}\}$ are identically distributed random variables with zero expectations, and the coefficients μ and β are numeric constants. All three equations represent *linear* regression models because all are linear in the to-be-estimated constants μ, β, and γ.

What then is a *nonlinear* regression? Here are two examples:

$Y = \beta\log(\gamma X)$, which is linear in β but nonlinear in the coefficient γ

and

$T = \beta\cos(t + \gamma)$, which also is linear in β but nonlinear in γ.

Suppose now, you've reason to believe there exists a cause-and-effect relationship between a *dependent variable Y*, and one or more other variables, X_1, X_2, ... often referred to as the *predictors*, that takes the linear form

$$EY = \mu + \beta_1 f_1[X_1] + \beta_2 f_2[X_2] + \dots$$

where EY is the expected value of Y for the given values of the predictors, and $f_1, f_2, \dots$ are specific functions of the predictors, for example $f_1[X_1] = X_1$, $f_2[X_2] = \log[X_2]$ and so forth.

The standard or ordinary least-squares method of solving for the coefficients μ, β_1, ... is to choose their values so as to minimize the sum of squares of the difference between the observed values of Y and the predicted values. Thankfully, computer software does it for us.

Three words of caution:

1. Computer software including R will attempt to find values for the coefficients μ, β_1, …even if there is *no* cause-and-effect relationship. We will assume in what follows that you use your brain as well as your software when you construct a model.
2. Don't confuse cause and effect. For example, an Orange County social worker reported to a CASA meeting that problems in the home environment often accompanied (rather than were accompanied by) a decline in a student's grade point average.
3. Be aware that still other unrepresented variables may be responsible for the changes in the values of both the dependent and independent variable.

An Example

Suppose that your physician keeps telling you to modify your diet. "Your systolic blood pressure (SBP) is a function of your age," he says. Let's look at the data:

```
> Age=c(39,47,45,47,65,46,67,42,67,56,64,
  56,59,34,42)
> SBP=c(144,220,138,145,162,142,170,124,
  158,154,162,150,140,110,128)
> glm(SBP~Age)

Call: glm(formula = SBP ~ Age)
Coefficients:
(Intercept)   Age
95.613      1.047
Degrees of Freedom: 14 Total (i.e. Null);
 13 Residual
Null Deviance: 8796
Residual Deviance: 6963 AIC: 140.7
```

We interpret these results as the linear relationship, $\hat{E}(SBP) = \hat{a} + \hat{b}\,Age = 95.6 + 1.04\,Age$.

Notice that when we report our results, we drop excess decimal places that convey a false impression of precision.

Let's examine the *residuals*, that is, the differences between the values our equation would predict and what individual values of systolic blood pressure were actually observed.

```
> residuals(glm(SBP~Age))
1. 7.537 2. 75.158 3. -4.747 4. 0.158
 5. -1.696 6. -1.7947 7. 4.2091 8.-15.605
 9. -7.7901 10.-0.269 11. -0.649 12. -4.269
 13.-17.411 14.-21.225 15. -11.605
```

Consider the fourth residual in the series, 0.15. This is the difference between what was observed, $SBP = 145$, and what our ordinary least squares (OLS) regression line estimates as the expected SBP for a 47-year-old individual, $E(SBP) = 95.6 + 1.04*47 = 144.8$. Note that we were furthest off the mark in our predictions, that is, we had the largest error residual, in the case of the youngest member of our sample, a 34-year-old.

Oops, I made a mistake; there is a still larger residual. Mistakes are easy to make when glancing over a series of numbers with many decimal places. You won't make a similar mistake if you first graph your results using the following R code (Figure 12.1):

```
> Pred = glm(SBP ~ Age)
> plot (Age, SBP)
> lines (Age,fitted(Pred)).
```

Now the systolic blood pressure value of 220 for the 47-year-old stands out from the rest.

Warning: If you don't see the graph at this point, it's because you've not been following along, entering the code in R.

Confidence Intervals

Let's take a closer look at the analysis underlying Figure 12.1:

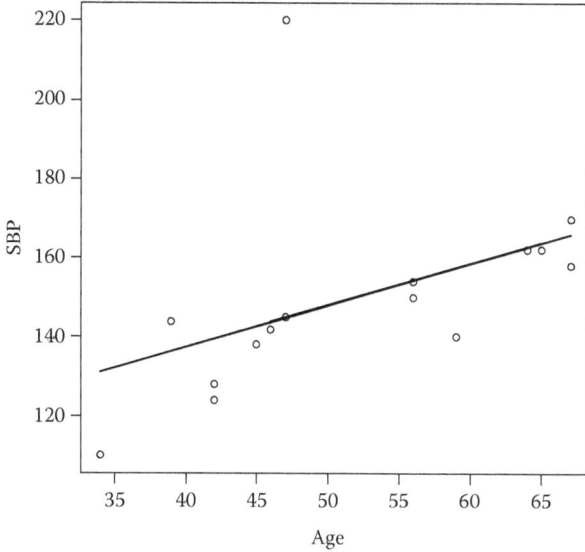

FIGURE 12.1 Plot of systolic blood pressure as a function of age.

```
> summary(glm(SBP~Age))

Call:
glm(formula = SBP ~ Age)

Deviance Residuals:
 Min 1Q Median 3Q Max
-14.5402 -4.8981 -0.1823 7.4230 12.7993

Coefficients:
   Estimate Std. Error t value Pr(>|t|)
(Intercept) 45.2152 16.5772 2.728 0.0197 *
Age        1.8303 0.3075 5.952 9.56e-05 ***
---
Signif. codes: 0 '***' 0.001 '**'
 0.01 '*' 0.05 '.' 0.1 ' ' 1

(Dispersion parameter for gaussian family
taken to be 69.50131)
```

While the effect of Age is said to be highly significant, the probability of a greater value of the test statistic being less than 0.001, this probability is based on the assumption that the residual errors are Gaussian (that is, that they come a normal distribution).

A normal distribution arises whenever the residual error, that part of the systolic blood pressure reading not explained by age, is the result of a large number of yet-to-be-determined factors, each of which makes only a small contribution to the total.

The same applies to the confidence intervals one can obtain with the aid of R's `confint` function.

```
> confint(glm(formula = SBP ~ Age))
```

Waiting for profiling to be done ...

```
 2.5 % 97.5 %
(Intercept) 12.724346 77.705959
Age 1.227596 2.432932
```

As was noted in Chapter 8 in the section on confidence intervals, this means that the probability is 90% that the interval 1.23 to 2.43 includes the slope of the regression line. This interval excludes zero, confirming that a statistically significant relationship exists between age and systolic blood pressure.

Confidence Bounds

We can provide confidence lines for our regression line with the following, not-entirely-obvious code:

```
> new = data.frame(Age)
> clim=as.data.frame(predict
  (pred,new,level=0.95,interval=
  "confidence"))
> plot(Age,SBP)
> abline(pred)
> new=sort(new)
```

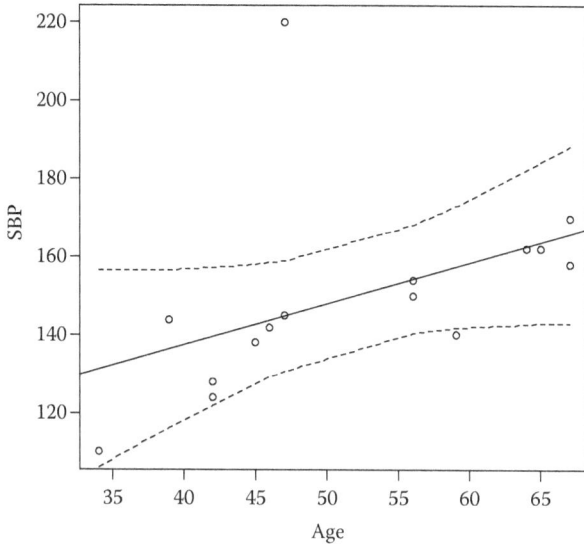

FIGURE 12.2 Ninety-five percent confidence bounds for predictions based on our regression line.

```
> lines(cbind(new,sort(clim$lwr)), lty=
  "dashed")
> lines(cbind(new,sort(clim$upr)), lty=
  "dashed")
```

The results are displayed in Figure 12.2.

Distribution-Free Intervals

Again, these results are based on the assumption that the residual errors are independent, identically distributed, and come from a normal distribution.

In some instances, it may be possible to find a function f of the independent variable, such that $f[Y] = \alpha X_1 + \beta X_2 + z$, where the residual z has a normal distribution. One might try to fit log[Y] when $E[Y] = X_1^{\alpha} \exp[-\beta X_2]$. Even then the residuals might or might not be Gaussian.

If we can assume only that the residual errors are independent, though not that they are identically distributed or Gaussian, we still can obtain confidence intervals by making use of the bootstrap procedure described in Chapter 8.

First, we collect the variables in a single data frame so as to sample each pair of related observations as a unit,

```
data=data.frame(SBP, Age)
N =400 #set number of bootstrap samples
n=length (SBP)
stat=numeric(N) #create a vector in which
 to store the results
for(i in 1:N){
# sample from the numbers 1 through
length(SBP)
ind=sample(n,n, replace=T)
boot= data[ind,]
stat[i]=glm(SBP~Age,data=boot)
}
quantile (stat,prob=c(.05,.95))
```

Improving the Fit

We can improve the fit of our model to the observations in any or all of five different ways:

- Diminishing the effects of extreme observations
- Eliminating misleading observations or outliers
- Narrowing the range over which the model will be applied
- Using higher-order polynomials
- Increasing the number of predictors

Diminishing the Effects of Outliers

We can diminish the effects of extreme observations on our model in much the same way as we diminished the effects of extreme observations on our calculation of a central value for a distribution, by minimizing the sum

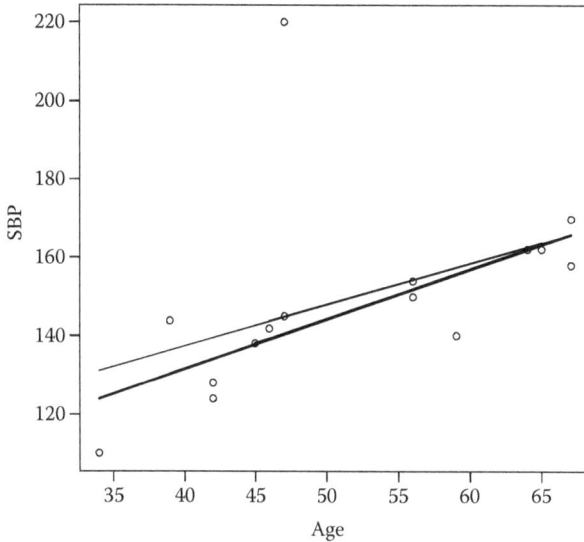

FIGURE 12.3 LAD regression diminishes the effects of outliers.

of the absolute values of the deviations of our observations about the regression line rather than the sum of the squared deviations. The technical details of the process are provided in the next chapter in the section on quantile regression. Figure 12.3 depicts the result.

Eliminating Outliers

Consider the effect on the regression line if we were to eliminate the systolic blood pressure reading of 220 for the 47-year-old recorded in the previous example. See Figure 12.4.

Narrowing the Range

In many physical and biological examples such as the volume of a gas viewed as a function of the applied pressure, or the frequency of cricket chirps as a function of temperature, the relationship may be linear for most of the predictor's range, but quickly turn nonlinear for extreme values of the predictor.

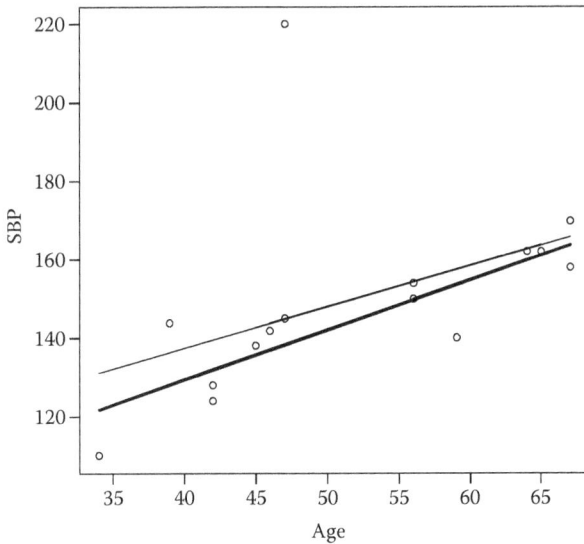

FIGURE 12.4 Effect on the model of eliminating an outlying observation.

For example, according to the inside of the cap on a bottle of Snapple's Mango Madness, "the number of times a cricket chirps in 15 seconds plus 37 will give you the current air temperature." Really? Ask yourself how many times one would expect to hear a cricket chirp in 15 seconds when the temperature is 35°F? Or 135°F? The answers are in a footnote.[*]

If you suspect nonlinear behavior—here, a graph often can be diagnostic—it would be best to break up the range of the predictor and develop a separate model for each portion.

Adding Higher-Order Terms to a Model

In Figure 12.5, we plotted a simulated data set *Y* as a function of a simulated data set *X*. Note that the best-fitting regression line appears to cut through the observations, leaving all the values in the middle-range of the predictor *X* lying below the line, much the way a straight line would cut across a parabola. This suggests that we

---

[*] At 35°F, the cricket is hibernating; at 135°F, we're talking about a pan-fried cricket.

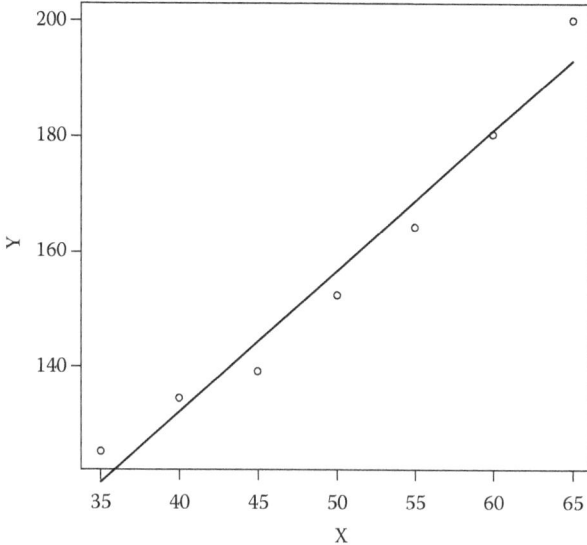

FIGURE 12.5 Plot of *Y* as a function of *X*.

might try to fit a higher-order polynomial such as *a* + *b* X + cX² to the data. The result is shown in Figure 12.6.*

A word of caution: We can get an increasingly better fit by incorporating higher-order terms in our model. A fifth-order polynomial would fit the six data points shown in the preceding figure perfectly. But the result may well be meaningless.

Goodness of Fit Does Not Guarantee Predictive Success

Increasing the Number of Predictors

In this example and the one following, we use functions of the observations as predictors. For example, if we had observations on body temperature taken at various times of the day, we would write

---

* Just a reminder: Although a parabola as far as the predictor is concerned, the polynomial is linear in it coefficients.

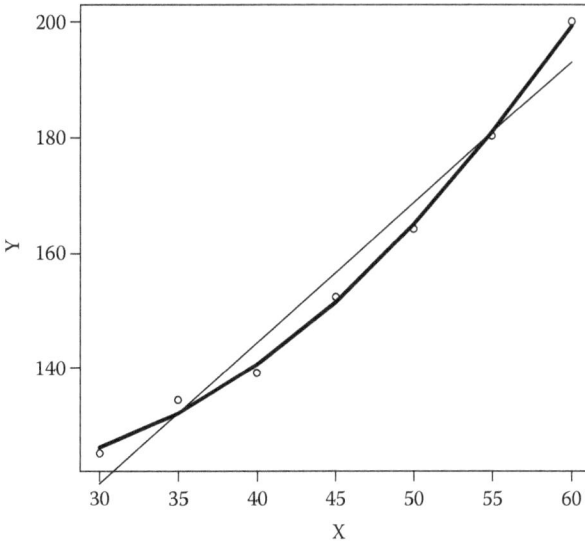

FIGURE 12.6 Plot of *Y* as a linear function of *X* and *X*².

```
> glm (temp ~ I(cos (2*pi*(time +
300)/1440)))
```

when we wanted to obtain a set of regression coefficients. Or, if we had data on weight and height as well as age and systolic blood pressure, we would write

```
> glm(SBP ~ Age+I(100*(weight/
(height*height)))).
```

If our regression model is $EY = \mu + \alpha X + \beta W$, that is, if we have two *predictor variables* X and W, we would write glm(Y~X+W). If our regression model is $EY = \mu + \alpha f[X] + \beta g[W]$, we would write glm(Y~f(X)+ g(W)), where the functions f and g have already been defined in R.

Note that the glm() function automatically provides an estimate of the *intercept* μ. The term intercept stands for "intercept with the Y axis," since, as can be seen from the equation, when $f[X]$ and $g[W]$ are both zero, the expected value of Y will be μ.

Sometimes, we know in advance that the intercept will be zero. Two good examples would be when we measure distance traveled as a function of time or the amount of substance produced in a chemical reaction. We can force glm() to set the intercept equal to zero by writing glm(Y~0+X).

Sometimes, we would like to introduce a cross product or interaction term into the model, $EY = \mu + \alpha X + \beta W + \gamma XW$, in which case, we would write glm(Y~X*W).

When used in expressing a linear model formula, the arithmetic operators in R have meanings quite distinct from their usual ones. Thus, in Y~X*W the "*" operator defines the regression model as $EY = \mu + \alpha X + \beta W + \gamma XW$ and not as $EY = \mu + \gamma X*W$ as you might expect. If we wanted the latter model, we would have to write Y~I(X*W). The I( ) function restores the usual meanings to the arithmetic operators.

In a further study of systolic blood pressure as a function of age, the height and weight of each individual were recorded. The latter were converted to a Quetelet index using the formula QUI = 100*weight/height^2. Suppose we were to fit a multivariate regression line of systolic blood pressure with respect to age and the Quetelet index using the following information:

```
Age = c(41,43,45,48,49,52,54,56,57,59,62,
  63,65)
SBP = c(122,120,135,132,130,148,146,138,
  135,166,152,170,164)
QUI = c(3.25,2.79,2.88,3.02, 3.10,3.77,
  2.98,3.67,3.17,3.88,3.96,4.13,4.01)

> summary(glm(SBP~Age+QUI))

Call:
glm(formula = SBP ~ Age + QUI)

Deviance Residuals:
    Min      1Q    Median      3Q      Max
```

```
-11.0857 -3.8031  -0.1754  6.6888  10.6540
Coefficients:
   Estimate Std. Error t value Pr(>|t|)
(Intercept) 35.4093 16.7275 2.117 0.0603 .
Age          1.1958 0.4932 2.425 0.0358 *
QUI 12.7281 8.0277 1.586 0.1439
---
Signif. codes: 0 '***' 0.001 '**' 0.01 '*'
  0.05 '.' 0.1 ' ' 1

(Dispersion parameter for gaussian family
 taken to be 61.0931)

Null deviance: 3226.92 on 12 degrees of
 freedom
Residual deviance: 610.93 on 10 degrees
 of freedom
AIC: 94.943
```

These results would suggest that systolic blood pressure does depend on Age, but not on the individual's Quetelet index. Of course, a larger sample might suggest the opposite (as described in Chapter 9).

Let us add an interaction term to our model. That is

```
E[SDP] = μ + αAge +βQUI + γAge*Qui

Call:
glm(formula = SBP ~ Age * QUI)

Deviance Residuals:
 Min 1Q Median 3Q Max
-9.715 -4.845 -1.930 5.104 11.157

Coefficients:
            Estimate   Std. Error  t value   Pr(>|t|)
(Intercept) 134.8581   160.0979    0.842     0.421
Age          -0.6288     2.9643   -0.212     0.837
QUI         -17.0368    48.3528   -0.352     0.733
Age:QUI       0.5381     0.8613    0.625     0.548
```

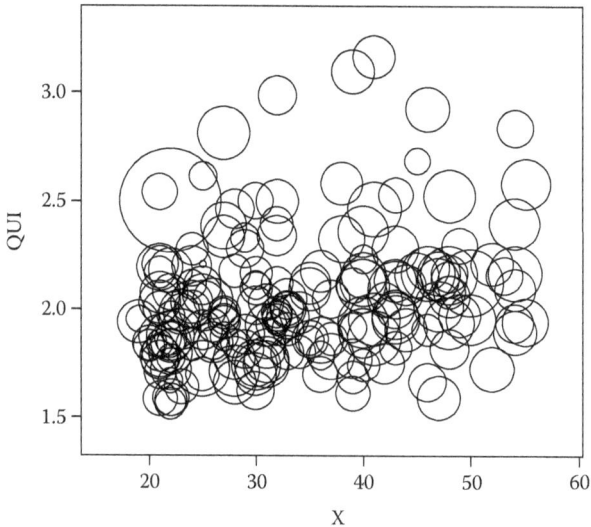

FIGURE 12.7   Cholesterol level as a function of age and the Quetlet index for the Werner blood chemistry data.

```
(Dispersion parameter for gaussian family
taken to be 65.05912)

 Null deviance: 3226.92 on 12 degrees of
    freedom
Residual deviance: 585.53 on 9 degrees
    of freedom
AIC: 96.39

Number of Fisher Scoring iterations: 2
```

How are we to interpret these latter results in which none of the individual coefficients are significantly different from zero? Let's bootstrap to find out how *reproducible* these results are; that is, let us take a series of bootstrap samples from the original sample to simulate what might happen if we were to take additional samples from the original population, compute the regression coefficients each time, and see if we obtain consistent results:

```
n = length(SBP)
#collect all variables in a single frame
 so as to sample each trio of related
 observations as a unit
data=data.frame(SBP,Age,QUI)
#set number of bootstrap samples
N =400
#create a matrix in which to store the
 results
stat=array(numeric(3*N),dim=c(3,N))
for(i in 1:N){
 ind=sample(n,n, replace=T)
 boot= data[ind,]
 y=glm(SBP~Age*QUI,data=boot)
 stat[1,i]=y[[1]][2]
 stat[2,i]=y[[1]][3]
 stat[3,i]=y[[1]][4]
 }
quantile(stat[1,],prob=c(0.05,0.95))
quantile(stat[2,],prob=c(0.05,0.95))
quantile(stat[3,],prob=c(0.05,0.95))
```

If you're like me, then the preceding code may strike you as utter nonsense. To see what is happening, bring up R on your computer and execute the following code in the order shown:

```
n = length(SBP)
data=data.frame(SBP,Age,QUI)
N =400
stat=array(numeric(3*N),dim=c(3,N))
ind=sample(n,n, replace=T)
boot= data[ind,]
data
boot
```

Notice that the vector boot holds a single sample with replacement taken from the subjects that were in the original data set. All the observations on a single subject are retained as a unit.

Now execute these commands to see how the program works:

```
y=glm(SBP~Age*QUI,data=boot)
stat[1,i]=y[[1]][2]
```

To obtain a bootstrap confidence interval for the coefficient of Age, run the complete program including the for loop and look at the results of `quantile(stat[1,],prob=c(0.05,0.95))`:

```
        5%          95%
 -7.000759   4.116497
```

Your results will be different from mine, since the results of taking bootstrap samples are strictly random. The important point is that the bootstrap confidence interval for the coefficient of Age includes 0. This suggests that with a slightly different sample from the same population, Age might not have been of any value in this model as a predictor.

Validation

The bootstrap is just one possible method of validating a model. A superior but far more expensive method is to take repeated samples from the population. A compromise, possible if the original sample is large enough, is to split the sample in two, using one part, termed the "training set," to develop the model, and the balance of the observations to test the model's goodness of fit.

Categorical Predictors

The linear regression model is a quantitative one. When we write $Y = 3 + 2*X$, we imply that the product $2*X$ will be meaningful. This will be the case if X is a metric variable. In many surveys, respondents use a nine-point *Likert scale*, where a value of "1" means they definitely disagree with a statement, and "9" means they definitely agree. Although such data are ordinal and not metric, the regression equation is still meaningful.

When one or more predictor variables are categorical, sex or race, for example, we must use a different approach. R makes it easy to accommodate categorical predictors. We first declare the categorical variable to be a factor. The regression model fit by the glm() function will include a different additive component for each level of the factor or categorical variable.

For example:

```
> Age = c(41,43,45,48,49,52,54,56,57,59,
  62,63,65)
> SBP = c(122,120,135,132,130,148,146,138,
  135,166,152,170,164)
sex = c("b",rep("g",3),"b",rep("g",5),
  rep("b",3))
sexf = factor(sex)
glm(SBP~Age*sexf)
```

Too Many Predictors

As the number of potential predictors grows, occasionally without reference to common sense—is the potential predictor really a potential contributor to the dependent variable's value?—there is increasing risk that the model's fit may be due to chance alone, and the results of our calculations applicable only to the data at hand.

In a carefully crafted set of simulations, David Freedman (1983) found a highly significant correlation as measured by R^2 between a dependent variable and a set of unrelated and totally independent predictors. As he notes in the introduction to the article, "If the number of variables is comparable to the number of data points, and if the variables are only imperfectly correlated among themselves, then a very modest search procedure will produce an equation with a relatively small number of explanatory variables, most of which come in with significant coefficients, and a highly significant R^2. Such an equation is produced even if Y is *totally unrelated* to the Xs."

He generated 5100 independent normally distributed random "observations." You can duplicate his efforts with the R command, `rnorm(5100)`. He divided these simulated observations into 51 sets of a hundred values each, with the first 50 sets corresponding to the values of 50 predictors, and the 51st to the values of the dependent variable.

Fifteen coefficients of the 50 were significant at the 25% level, and one was significant at the 5% level.

Today with the help of R, one might do a similar selection, either by using stepwise regression as in

```
library(MASS)
fit <- lm(y~x1+x2+x3+... x50,data=mydata)
step <- stepAIC(fit, direction="both")
step$anova # display results
```

or by all subsets regression as in

```
library(leaps)
attach(mydata)
leaps<-regsubsets(y~x1+x2+x3+...+ x50,
 data=mydata, nbest=15)
# view results
summary(leaps)
# plot a table of models showing variables
 in each model.
# models are ordered by the selection
 statistic.
plot(leaps,scale="r2")
# plot statistic by subset size
library(car)
subsets(leaps, statistic="rsq")
```

Friedman constructed a second model by the same method using only the 15 variables that had proven significant on the first pass. The resulting model had an R^2 of 0.36. The model coefficients of six of the (completely unrelated) predictors were significant at the 5% level! Can one ever be sure that the statistically significant variables we uncover in our research are truly explanatory or are merely the result of chance?

A partial answer may be found in an article by Gail Gong (1986). She constructed a logistic regression model using the observations Peter Gregory had made on 155 chronic hepatitis patients in the Stanford Medical Center, 33 of whom died. Using 19 predictors derived from medical histories, physical examinations, x-rays, liver function tests, and biopsies, the pair hoped to be able to identify patients at high risk.

Gong's logistic regression models were constructed in two stages. In the first stage, each of the potential predictors was evaluated on a predictor-by-predictor basis. Thirteen proved significant at the 5% level when applied to the original data. A forward multiple regression was applied to these 13 variables, and four were selected for use in the predictor equation.

At this point, she applied the bootstrap procedure we described earlier under "Validation." The R^2 values of the final models associated with each bootstrap sample taken from the 155 patients varied widely. Two of the original four predictors always appeared in the final models, and five other variables were incorporated in *some* but not all of the models.

Validate your models. Goodness of fit is not prediction, and the bootstrap permits one to simulate what might occur if one made additional observations.

What about Missing Data?

When we have a large number of variables, it is highly likely that one or more of them may be missing. Our first step, as noted in Chapter 6, is to see if they are missing in clusters, and if so to perform subsampling to remedy the situation. Or, if the missing data is primarily associated with one or two variables, simply eliminate those variables from further consideration.

Thereafter, we may either delete any observations with missing values (which may leave us with very little data indeed) or impute values to the missing observations (See Harrel, 2001 on methods of imputation).

Analysis of Variance

This might seem an odd chapter to locate a method of testing hypotheses, until one realizes that the analysis of variance is also a method of regression in which the values of the predictors occur at a small number of fixed values. The distinction from a regression equation lies in how we display the results.

Begin by loading a data set which records the relationship between crop yield and the amount of sunlight and fertilizer with which the plants were supplied.

```
> data=read.table("http://statcourse.com/
  research/crop.csv",sep=",", header=TRUE)
> edit (data)
> attach(data)
> pred=lm(Yeld~Sun*Fert)
> anova(pred)
Analysis of Variance Table
Response: Yeld
         Df  Sum Sq  Mean Sq  F value     Pr(>F)
Sun       1  997.56   997.56   43.902  2.444e-05 ***
Fert      2 2952.44  1476.22   64.968  3.652e-07 ***
Sun:Fert  2  589.78   294.89   12.978  0.0009986 ***
Residuals 12 272.67 22.72
Signif. codes: 0 '***' 0.001 '**' 0.01 '*' 0.05 '.'
               0.1 ' ' 1
```

Note our use of the `lm()` function rather than `glm()` and of `anova()` rather than `summary()`.

Nested Models

Three casks selected at random from each delivery of a chemical paste were sampled at random and two analytical tests carried out on each sample to obtain the results reported by Davies and Goldsmith (1972, Table 6.5, p. 138).

To obtain their data, install, and then load the library `lme4`. We make use of a slight variant of the anova command to obtain results for the nested design:

```
> edit (data)
> fit=lm(strength~batch/cask,Pastes)
> anova(fit)
Response: strength
            Df  Sum Sq  Mean Sq F value    Pr(>F)
batch        9  247.40   27.489  40.544 2.280e-14 ***
batch:cask  20  350.91   17.545  25.878 9.791e-14 ***
```

Summary

In this chapter, you were provided with the R code for developing linear regression models utilizing ordinary least squares. You were provided with five different ways for improving the fit of your models.

As a model developed on one set of data may fail to fit a second independent sample nearly as well, you were shown how to use the bootstrap to validate your models, as well as to select the important independent variables when you have large number of potential predictors.

Finally, you were shown that the results of an analysis of variance are closely related to those of ordinary least squares linear regression.

To Learn More

Good and Hardin (2008) consider common errors in model building. Friedman, Furberg, and DeMets (1998) cite a number of examples of clinical trials using misleading surrogate variables. Mielke et al. (1996) investigates the effects of sample size, type of regression model, and noise to signal ratio on the decrease or shrinkage in fit from the data used to first develop a model to the future data to which one might apply it.

Chapter 13

Alternate Regression Methods

Prescription

In this chapter, we consider a large number of alternatives to ordinary least-squares regression:

If the outcome variable is continuous:

Least absolute deviation (LAD) regression should be used in preference to ordinary least squares (OLS) in four circumstances:

1. To reduce the influence of outliers
2. If the losses associated with errors in prediction are additive, rather than large errors being substantially more important than small ones
3. If the conditional distribution of the dependent variable given the predictors is not symmetric and we wish to estimate its median rather than its mean value
4. If the conditional distribution of the dependent variable is heavy in the tails

Errors-in-variables (EIV) or Deming regression should be used whenever the values of the predictors cannot be specified precisely, such as when making comparisons between old and new methods of measurement.

If the relation between the outcome variable Y and the predictors $X = \{X_1, X_2, \ldots X_n\}$ requires a link function L in order to take the linear form $L[Y] = AX + b + z$, use general estimating equations. Two examples are *logistic regression* for use when the outcome variable is binary or dichotomous and *Poisson regression* when the outcome is a count.

The definition of a linear regression model advanced in the previous chapter is applicable to all these modeling methods, as are the R functions `glm()`, `coef()`, `residuals()`, and `predict()`. Our cautions in the previous chapter regarding extrapolating beyond the observed ranges of the predictors and of employing excessive numbers of terms in the model apply to these models as well.

A final section of this chapter deals with classification.

LAD Regression

Least absolute deviation (LAD) regression is used to predict the median value of a variable which depends in a linear fashion upon a set of predictors. That is, when $Y = AX + b + z$, where b is a constant and z a random variable.

Least Absolute Deviation regression (LAD) attempts to correct one of the major flaws of ordinary least squares regression (OLS), that of sometimes giving excessive weight to extreme values. The LAD method solves for the coefficients in the regression equation by finding those values for which the sum of the *absolute deviations* of the observed about the predicted values is a minimum. A deviation that is twice as large as another contributes twice as much to the solution, not four times as much. The real difference can be seen when fitting a set of values such as those shown in Figure 13.1.

Recall from Chapter 8 that the sum of the absolute values of the deviation of the observations about the median or 50th percentile is a minimum. Finding the corresponding values of the coefficients in the matrix A

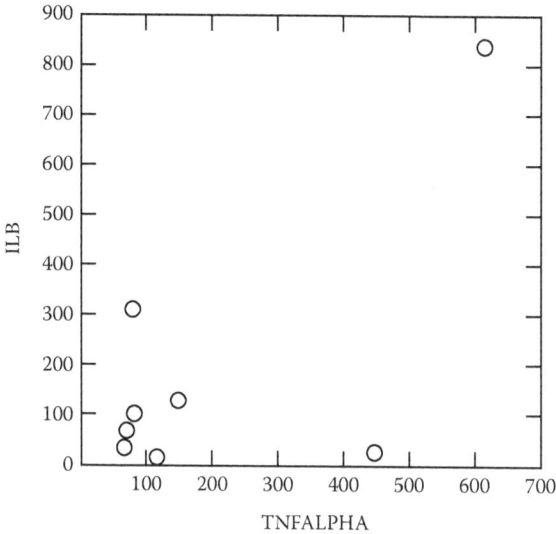

FIGURE 13.1 A single outlying value can radically influence the values of OLS coefficients.

to obtain this minimum is computationally demanding, as it requires linear programming. The use of a computer is essential, but who does hand calculations these days?

You need to install a new package in R in order to do the calculations. The easiest way to install it is to get connected to the Internet, pull down the Packages menu, select "load packages," and then quantreg.

The installation, which includes downloading, unzipping, and integrating the new routines, is done automatically. It needs to be done once and once only. But each time you load R into memory, before you can use the LAD routine, you'll need to load the supporting functions into computer memory by typing

```
library(quantreg)
```

Let us look a second time at the data sets

```
Age=c(39,47,45,47,65,46,67,42,67,56,64,
    56,59,34,42)
```

```
SBP=c(144,220,138,145,162,142,170,124,
   158,154,162,150,140,110,128)
```

To perform the LAD regression, type

```
rq(SBP ~ Age)
```

and obtain the following output:

```
Coefficients:
(Intercept)          Age
  81.157895   1.263158
```

To display a graph, type

```
f = coef(rq(SBP ~ Age))
pred = f[1] + f[2]*Age
plot (Age, SBP)
lines (Age, pred)
```

In Figure 13.2, we compare a plot of the predicted median (the thicker, darker line) with that of the predicted mean. We may obtain the identical LAD regression line with the following R code:

```
g = rq(SBP ~ Age)
  plot (Age, SBP)
  lines(Age,predict(g))
```

Quantile Regression

Often one's interest will be in quantiles of the population other than the median. Economists and welfare workers will want to predict the number of individuals whose incomes place them below the poverty line; physicians, bacteriologists, and public health officers will want to estimate the proportion of bacteria that will remain untouched by various doses of an antibiotic; ecologists and

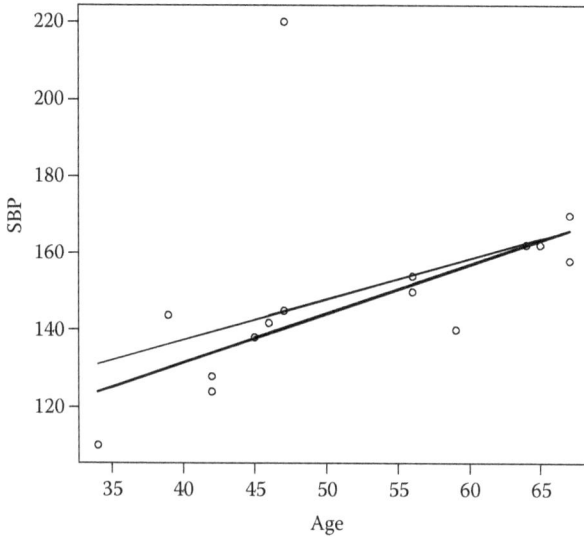

FIGURE 13.2 Regression lines of predicted mean (light) and median (dark).

nature lovers will want to estimate the number of species that might perish in a toxic waste spill; and industrialists and retailers, car manufacturers and dealers, for example, will need to know what proportion of the population might be interested in and could afford their new product.

A simple modification of our previous effort will produce the desired results (Figure 13.3):

```
> rq(SBP~Age,tau=c(0.1,0.9))

Call:
rq(formula = SBP ~ Age, tau = c(0.1, 0.9))
Coefficients:

            tau = 0.1 tau = 0.9

(Intercept)  60.545455   107.7857143
Age           1.454545     0.9285714
Degrees of freedom: 15 total; 13 residual
```

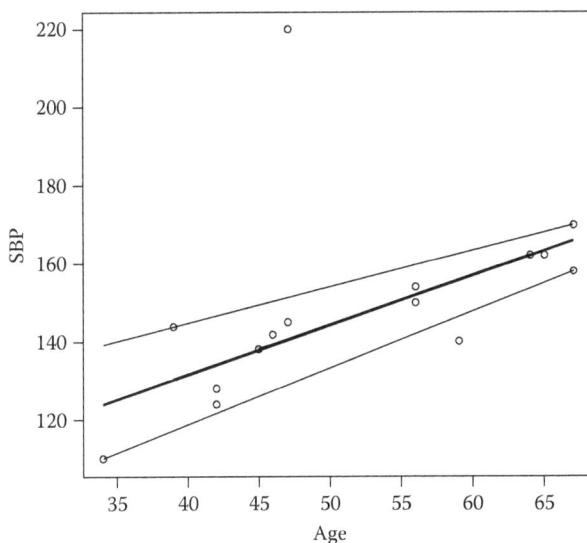

FIGURE 13.3   Quantile regression lines for P_{10}, P_{50}, and P_{90}.

Errors-in-Variables Regression

The need for errors-in-variables (EIV) or Deming regression is best illustrated by the struggles of a small medical device firm to bring its product to market. Their first challenge was to convince regulators that their long-lasting device provided equivalent results to those of a less-efficient device already on the market. In other words, they needed to show that the values V recorded by their device bore a linear relation to the values W recorded by their competitor, that is, that $E[V] = a + bW$.

In contrast to the examples of regression we looked at earlier, the errors inherent in measuring W (the so-called predictor) were as large if not larger than the variation inherent in the output V of the new device.

The EIV regression method they used to demonstrate equivalence differs in two respects from that of OLS:

1. With OLS, we are trying to minimize the sum of squares $\Sigma(y_{oi} - y_{pi})^2$ where y_{oi} is the ith observed value of Y and y_{pi} is the ith predicted value. With EIV, we are trying to minimize the sums of squares of errors, going both ways: $\Sigma(y_{oi} - y_{pi})^2/VarY + \Sigma(x_{oi} - x_{pi})^2/VarX$.
2. The coefficients of the EIV regression line depend on the ratio lamda of the variances of X and Y.

To compute the EIV regression line, you'll need an estimate of this ratio (usually obtained by taking a set of duplicate measurements) along with the following R function:

```
eiv = function (X,Y, lamda){
  Sxx = var(X)
  Syy = var(Y)
  Sxy = cov(X,Y)
  diff = lamda*Syy-Sxx
  root = sqrt(diff*diff+4*lamda*Sxy*Sxy)
  b = (diff + root)/(2*lamda*Sxy)
  pred=mean(Y) +b*(X-mean(X))
  return (pred)
}
```

Generalized Linear Models

The generalized linear model (GLM) generalizes linear regression by allowing the linear model to be related to the response variable via a link function L. Additionally, the variance of each output variable as well as its mean may be a function of its predicted value. In symbols, $L[EY] = $ AX and $\text{Var}(Y) = f[$ AX $]$. Two examples are considered in what follows, Poisson regression and logistic regression.

Poisson Regression

Poisson regression is appropriate when the dependent variable is a count, as is the case with the arrival of individuals in an emergency room (see Figure 13.4). It is also applicable to the spatial distributions of tornadoes and of clusters of galaxies. To be applicable the events underlying the outcomes must be independent in the sense that the occurrence of one event will not make the occurrence of a second event in a nonoverlapping interval of time or space any more or less likely. This model takes the form $\log \left[E Y \right] = AX + b + z$.

As an example, while working in R, upload a set of data relating the number of super-clusters of galaxies to their distance from the sun:

```
> data =read.table("http://statcourse.
  com/research/sclust.csv", sep=",",
  header = TRUE)
```

Attach the data file so we can reference the variables individually:

```
> attach(data)
```

Analyze the data with the statement

```
> mod=glm(Number~Distance,family="poisson")
```

Note the need to specify that family = "poisson".

```
> f = coef(mod)
> pred=f[1]+f[2]*Distance
> pred=exp(pred)
> plot (Distance, Number)
> lines(Distance, pred)
```

As a second example, upload the simulated dataset created by the UCLA: Academic Technology Services, Statistical Consulting Group

```
> data= read.table("http://www.ats.ucla.
  edu/stat/R/dae/poisson_sim.csv",
  sep=",", header = TRUE)
> attach(data)
```

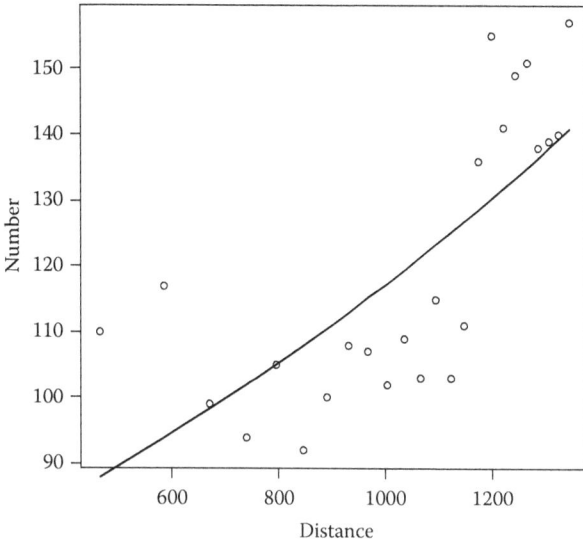

FIGURE 13.4 Poisson regression line.

Create labeled categories:

```
> pcat = factor(prog, labels=c("General",
"Academic", "Vocational"))
```

Analyze the data:

```
> hold=glm(num_awards ~ relevel(pcat,
"General") + math, family="poisson", p)
```

The chief differences from our prior efforts are twofold:

i. we specify family="poisson",
ii. we establish "General" as our reference group

Logistic Regression

Logistic regression is a standard technique in both pricing and survival analysis and is employed in many other areas as well. Once upon a time, I worked for a company whose software promised to guide other companies in making optimal bids, that is, a bid that would be lower than that submitted by other firms, yet high enough to provide the winning bidder with a substantial profit. The initial result of my company's analysis was a curve relating the dollar amount of the bid to its probability of success. As with any other probability, the probability of success p was such that $0 \leq p \leq 1$, although the outcome of the bid was inevitably success, 1, or failure, 0. The simplest of our models was $logit[p] = log[p/(1 - p)] = \mu + \alpha\$ + z$, where $\$$ represented the dollar value of the bid.

Of course, many other factors are involved in the award of a bid. The full logistic model takes the form $logit[p] = AX + b + z$, where X is a vector of predictor values, A is a vector of to-be-estimated coefficients, and z is a random variable.

Let's apply logistic regression to some data of Hosmer and Lemeshow (2000) on the presence or absence of coronary heart disease in individuals of various ages.

First, we enter the data

```
CHD=c(rep(0,4),1,rep(0,10),1,rep(0,6),1,
  rep(0,5),1,rep(0,2),1,rep(0,4),1,0,1)
CHD=c(CHD,rep(0,5),1,rep(0,2),1,rep(0,2),
  1,1,0,1,0,1,rep(0,2),1,0,1,1,
  rep(0,2),1,0,1)
CHD=c(CHD,rep(0,2),1,rep(1,3), 0,rep(1,5),
  0,0,rep(1,4),0,rep(1,4),0,rep(1,5),0,
  rep(1,3))
Age=c(20,23,24,25,25,26,26,28,28,29,rep
  (30,6),32,32,33,33,rep(34,5))
Age=c(Age,35,35,rep(36,3),rep(37,3),38,38,
  39,39,40,40,41,41,rep(42,4))
Age=c(Age,rep(43,3),rep(44,4),45,45,46,46,
  rep(47,3),rep(48,3),rep(49,3))
Age=c(Age,50,50,51,52,52,53,53,54,
  rep(55,3),rep(56,3),rep(57,6),rep(58,3))
Age=c(Age,59,59,60,60,61,62,62,63,64,64,
  65,69)
```

```
> pred=glm(CHD~Age,family =
  binomial("logit"))
> summary(pred)

Call:
glm(formula = CHD ~ Age, family =
 binomial("logit"))

Deviance Residuals:
    Min      1Q   Median      3Q      Max
-1.9718  -0.8456  -0.4576  0.8253   2.2859

Coefficients:
            Estimate Std. Error z value Pr(>|z|)
(Intercept) -5.30945    1.13365  -4.683 2.82e-06 ***
Age          0.11092    0.02406   4.610 4.02e-06 ***
Signif. codes: 0 '***' 0.001 '**' 0.01 '*' 0.05 '.'
               0.1 ' ' 1
```

(Dispersion parameter for binomial family taken to be 1)
Null deviance: 136.66 on 99 degrees of freedom
Residual deviance: 107.35 on 98 degrees of freedom
AIC: 111.35
Number of Fisher Scoring iterations: 4

As with OLS, we can obtain confidence intervals for the coefficients:

```
confint(pred,level=0.8)
```

Waiting for profiling to be done:

```
                 10%          90%
(Intercept)  -6.84273444  -3.9269664
Age           0.08149468   0.1433769
```

Then the code:

```
> plot(Age,CHD)
> lines(Age,fitted.values(pred))
```

yields Figure 13.5.

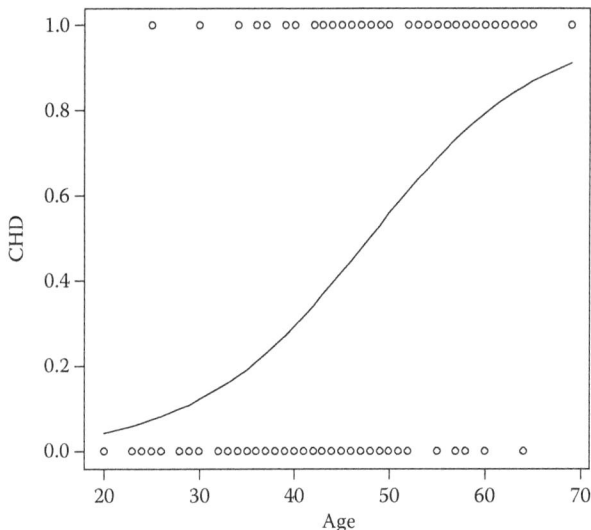

FIGURE 13.5 Increasing probability of coronary heart disease with age.

Note that the R command lines (Age,fitted. values(pred)) is equivalent to performing all of the following commands:

```
> f=coef(pred)
> probit=f[1]+f[2]*Age
> tmp=exp(probit)
> p=tmp/(1+tmp)
> lines(Age,p)
```

Classification

When we wish to classify an item (such as a mushroom) or an individual (such as a hospital patient) into one of three or more categories, we need to take a somewhat different approach. Begin by installing the Rweka package, then loading it into memory with the R command, library(Rweka).

Let's use the data recorded by Anderson (1935) of the sepal length, sepal width, petal length, petal width, and species of 150 iris plants as our training set:

```
> data =read.table("http://statcourse.com/
  research/Iris.csv", sep=",", header =
  TRUE)
> #examine the data to ensure it was
  loaded correctly
> edit(data)
> attach(data)
> species=Logistic
  (Class~SepLth+SepWth+PetLth+PetWth)
> summary(species)

=== Summary ===

Correctly Classified Instances  148  98.6667 %
Incorrectly Classified Instances  2   1.3333 %
Kappa statistic                 0.98
Mean absolute error             0.0164
Root mean squared error         0.0915
Relative absolute error          3.6884 %
Root relative squared error    19.4012 %
Coverage of cases (0.95 level)    100 %
Mean rel. region size (0.95 level) 36 %
Total Number of Instances         150

=== Confusion Matrix ===

a b c <-- classified as
50 0 0 | a = setosa
0 49 1 | b = versicolor
0 1 49 | c = virginica
```

To apply our classification scheme to a new specimen, create a data frame to hold the test data:

```
> test
  SepLth SepWth PetLth PetWth
  1  5.5   3    4    2
> predict(species,newdata=test)
  [1] versicolor
  Levels: setosa versicolor virginica
```

Note that when `predict()` is applied to the output of the logistic function, it performs a quite different set of operations than when it is applied to the output of the glm function.

Modeling Survival Data

A counting process is a stochastic process starting at 0 (everyone alive, no units failed), which proceeds in unit steps as units fail or individuals relapse or leave the study. Let T^* denote the time that elapses before a unit fails (or a patient relapses). The survival function $S[t]$ is defined as the probability that T^* is greater than t. The hazard function $\lambda[t]$ is defined as the probability that an individual will die at almost exactly time t.

The objective of our models is to estimate these two functions.

Installing the R survival library provides access to the prognostic data for 228 patients suffering from advanced lung cancer. We can display the data simply by entering the following commands (Figure 13.6):

```
> library(survival)
> lung
```

The following R commands result in the survival curve and 95% confidence bounds displayed in Figure 13.6.

```
> kfit=survfit(Surv(time,status) ~1,
    data=lung, type="kaplan-meier")
> plot(kfit, mark.time=F, xscale=365.25,
    xlab="Years", ylab="Survival")
```

Setting `mark.time` = `F` or `FALSE` suppresses the appearance of markers that show when censoring occurred.

Setting `xscale` = `365.25` converts days to years.

Most often, we want to know which covariates will prove most valuable in predicting survival time. To do this, we

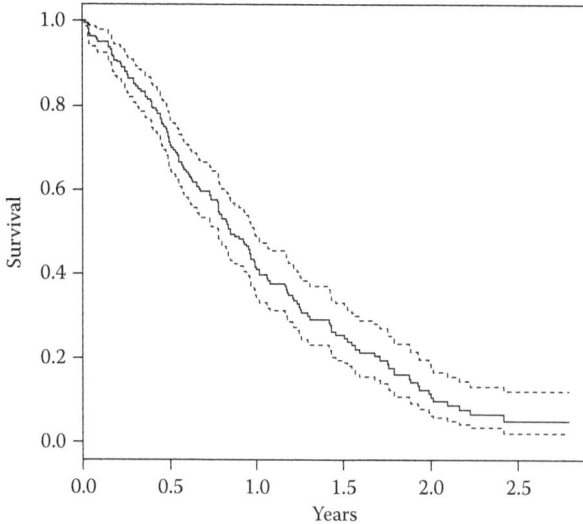

FIGURE 13.6 Estimates survival time for patients with advanced lung cancer. Dotted lines represent 95% confidence bounds.

need to specify the hazard as a function of the covariates. A frequent (though not always appropriate) choice is the Cox model, which specifies the hazard for an individual in the form $\lambda[t] = \lambda_0[t]\exp[\beta X[t]]$. Our objective is to estimate the vector of coefficients β.

Let us attempt this using the primary biliary cirrhosis of the liver or PBC data set pbc that is also part of the R survivor package:

```
> fit.pbc=coxph(Surv(time, status==2) ~
  age + sex +edema +log (bili) +
  log(albumin) + chol, data=pbc)
> print(fit,pbc)

Call:

coxph(formula = Surv(time, status == 2) ~
  age + sex + edema + log(bili) +
  log(albumin) + chol, data = pbc)
```

	coef	exp(coef)	se(coef)	z	p
age	3.22e-02	1.0328	0.009726	3.314	9.2e-04
sexf	-3.65e-01	0.6944	0.265021	-1.376	1.7e-01
edema	1.06e+00	2.8911	0.337563	3.145	1.7e-03
log(bili)	9.01e-01	2.4613	0.119048	7.566	3.9e-14
log(albumin)	-2.97e+00	0.0515	0.800550	-3.705	2.1e-04
chol	-6.97e-05	0.9999	0.000428	-0.163	8.7e-01

Likelihood ratio test = 165 on 6 df, p = 0 n = 284 (134 observations deleted due to missing values)

Neither edema nor sex appears to be of significance; not unexpectedly, the level of bilirubin is highly significant.

Principal Component Analysis

It's important to keep questionnaires as short as possible, yet to have some seemingly redundant questions for the purpose of validating responses. Assuming you've already conducted a pilot study, principal component analysis (PCA) is a method for rapidly assessing the associations among many different questions. In effect, PCA reduces a hard-to-read matrix of correlations to a two-dimensional display.

Summary

In this chapter, you learned a variety of alternatives to ordinary least squares regression including quantile regression, errors-in-variable regression, and general linear models. We also considered counting models for time-to-event data. Regardless of the model employed, you should always attempt to validate the results, either by dividing the original data into two sets, a training set and a test set, or by bootstrapping.

To Learn More

For more on quantile regression, download Blossom and its accompanying manual from http://www.fort.usgs.gov/products/software/blossom/.

Stöckl, Dewitte, and Thierpont (1998) compare ordinary linear regression, Deming regression, standardized principal component analysis, and Passing–Bablok regression. Only iteratively reweighed general Deming regression produces statistically unbiased estimates of systematic bias and reliable confidence intervals of bias. For details of the recommended technique, see Martin (2000).

For insight into the theory and application of general linear modeling, see Hardin and Hilbe (2012). For more on logistic regression, particularly as it applies to survival data, see Hosmer et al. (2008) and Hilbe (2009). For more on modeling survival data, see Therneau and Grambsch (2000). To analyze higher-order contingency tables via log-linear models, see Agresti (2002). For more on modeling count models using R, see Hilbe (2011).

CHAPTER **14**

Decision Trees

Prescription

Use decision trees for both classification and regression. Use them whenever you need to make your findings intelligible to a non-statistician—they're much to be preferred when you need to testify in court. They automatically account for higher-order interactions. Decision trees make it easy to evaluate the relative importance of potential predictors. They can be modified to account for the relative costs of misclassification, for the prior probabilities of the various classes when these are known, and for the relative costs of obtaining the values of the various predictors.

Decision Trees versus Regression

In this section, we develop two models for predicting home values, the first by regression means, the second via a decision tree. The data consists of the median value (MEDV) of owner-occupied homes in about 500 U.S. census tracts in the Boston area in the 1970s, along with several potential predictors, some of which are continuous and some categorical:

CRIM	Per capita crime rate by town
ZN	Proportion of residential land zoned for lots over 25,000 ft²
INDUS	Proportion of nonretail business acres per town
CHAS	Charles River dummy variable (= 1 if tract bounds river; 0 otherwise)
NOX	Nitric oxides concentration (parts per 10 million)
RM	Average number of rooms per dwelling
AGE	Proportion of owner-occupied units built prior to 1940
DIS	Weighted distances to five Boston employment centers
RAD	Index of accessibility to radial highways
TAX	Full-value property tax rate per $10,000
PTRATIO	Pupil–teacher ratio by town
LSTAT	Percent lower status of the population

Using R to obtain the desired results via stepwise regression:

```
> data=read.table("http://statcourse.
  com/research/boston.csv", , sep=",",
  header = TRUE)
> summary(step(glm(MEDV~ CRIM+ZN+INDUS+CHAS
  +NOX+RM+AGE+DIS+RAD+TAX+
  PTRATIO+B+LSTAT)))
```

and fitting with respect to the AIC criterion yields the model:

Step: AIC = 3024.8
MV = 36.3 −0.11CRIM + 0.05ZN + 2.7CHAS + 17.38NOX + 3.80RM + 1.49DIS + 0.30RAD − 0.01 TAX + 0.94PT + 0.009B−0.52LSTAT

With Multiple R-Squared: 0.7406, Adjusted R-squared: 0.7348

Using R to develop a tree,

```
> data=read.table("http://statcourse.com/
  research/boston.csv", , sep=",", header
  = TRUE)
> library(rpart)
> attach(data)
> fit=rpart (MEDV~ CRIM+ZN+INDUS+CHAS+NOX+
  RM+AGE+DIS+RAD+TAX+ > PTRATIO+B+LSTAT)
> print(fit)
   1) root 506 42716.3000 22.53281
   2) RM< 6.92 429 17272.7800 19.91818
   4) LSTAT>=14.4 175 3373.2510 14.95600
   8) CRIM>=6.99 74 1085.9050 11.97838 *
   9) CRIM< 6.99 101 1150.5370 17.13762 *
   5) LSTAT< 14.4 254 6621.6120 23.33701
  10) DIS>=1.55 247 3644.9160 22.92146
  20) RM< 6.545 193 1589.8140 21.65648 *
  21) RM>=6.545 54 642.4720 27.44259 *
  11) DIS< 1.55 7 1429.0200 38.00000 *
   3) RM>=6.92 77 6171.1200 37.10000
   6) RM< 7.435 47 1929.3590 31.99574
  12) LSTAT>=9.65 7 432.9971 23.05714 *
  13) LSTAT< 9.65 40 839.1960 33.56000 *
   7) RM>=7.435 30 1098.8500 45.09667 *
```

In the preceding listing, starred items correspond to terminal nodes. For example, line 8 tells us that if RM (average number of rooms per dwelling) is less than 6.92, and if LSTAT (percent lower status of the population) is greater than or equal to 14.4, and if CRIM (the town's per capita crime rate) is greater than or equal to 6.99, the median value of owner-occupied homes in the area will be $14,956 (at least, it would be if this were still the 1970s).

Executing the R commands:

```
> plot(fit, uniform=TRUE, compress=TRUE)
> text(fit, use.n=TRUE)
```

yields Figure 14.1.

RM<6.941

LSTAT>=14.4   RM<7.437

CRIM>=6.992   DIS>=1.551   LSTAT>=9.65
                                                45.1
                                                n=30

                      RM<=6.543
11.98   17.14   38   23.06   33.74
n=74   n=101   n=7   n=7   n=39

              21.66   27.43
              193   55

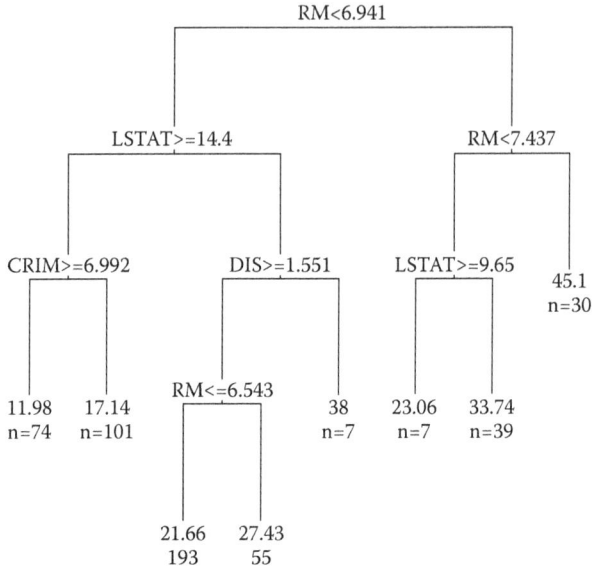

FIGURE 14.1 Decision tree to forecast median home values. Here, *n* denotes the number of homes in the training set assigned to a category.

How Trees Are Grown

Classification and regression tree builders make binary splits based on either yes/no (Sex = Male?) or less than/ greater than (RM < 6.92) criteria. The tree is built a level at a time beginning at the root. At each level, the computationally intensive algorithm due to Breiman et al. (1984) examines each of the possible predictors and chooses the predictor and the split value (if it is continuous) which best separates the data into more homogenous subsets.

This method is illustrated in Figures 14.2a, b and 14.3 using the Iris data set at http://statcourse.com/research/ Iris.csv.

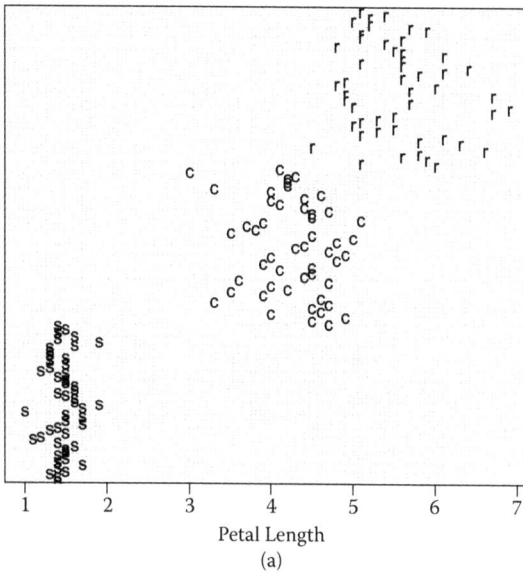

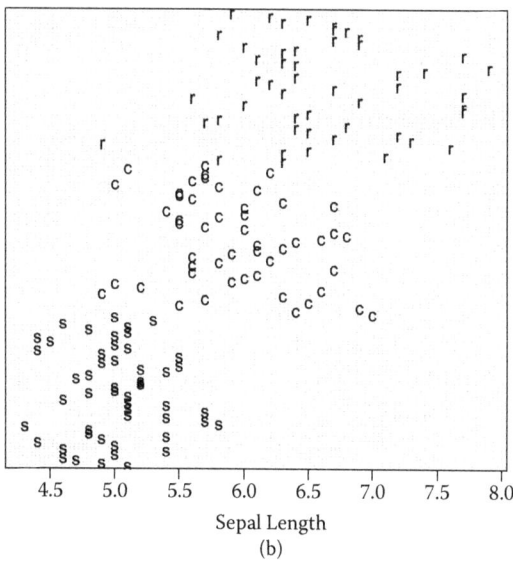

FIGURE 14.2 Subdividing Anderson's Iris data based on the values of a single variable.

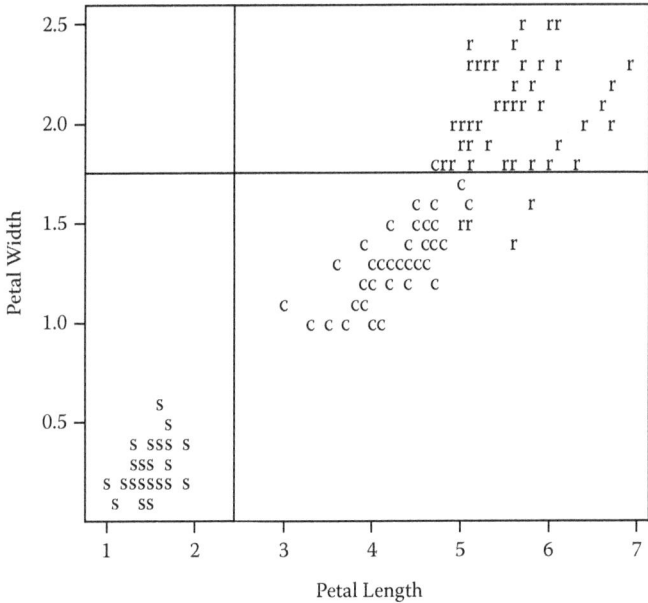

FIGURE 14.3   Subdividing Anderson's Iris data based on the values of two variables.

The best-fitting decision tree is

```
> rpart(Class~SepLth + PetLth +
  SepWth +PetWth)
node), split, n, loss, yval, (yprob)
 * denotes terminal node
1) root 150 100 setosa
   (0.33333333 0.33333333 0.33333333)
2) PetLth< 2.45 50 0 setosa
   (1.00000000 0.00000000 0.00000000) *
3) PetLth>=2.45 100 50 versicolor
   (0.00000000 0.50000000 0.50000000)
6) PetWth< 1.75 54 5 versicolor
   (0.00000000 0.90740741 0.09259259) *
7) PetWth>=1.75 46 1 virginica
   (0.00000000 0.02173913 0.97826087) *
```

Sepal length and width play no part here.

Incorporating Existing Knowledge

In contrast to regression, one can incorporate one's knowledge of prior probabilities and relative costs into models built with a classification and regression tree. In particular, one can account for all of the following:

1. The prior probabilities of the various classifications.
2. The relative costs of misclassification
3. The relative costs of data collection for the individual predictors

Prior Probabilities

Although equal numbers of the three species of Iris were to be found in the sample studied by Anderson, it is highly unlikely that you would find such a distribution in an area where near where you live. Specifying that the setosa are the most plentiful yields a quite different decision tree:

```
> rpart(Class~SepLth + PetLth + SepWth
  +PetWth, parms=list
  (prior=c(.20,.20,.60)))
1) root 150 60.0 virginica
   (0.20000000 0.20000000 0.60000000)
2) PetLth< 4.75 95 28.2 setosa
   (0.51546392 0.45360825 0.03092784)
4) PetLth< 2.45 50 0.0 setosa
   (1.00000000 0.00000000 0.00000000) *
5) PetLth>=2.45 45 1.8 versicolor
   (0.00000000 0.93617021 0.06382979) *
3) PetLth>=4.75 55 3.6 virginica
   (0.00000000 0.03921569 0.96078431) *
```

Misclassification Costs

Suppose instead of three species of Iris, we had data on three species of mushroom: the *Amanita calyptroderma* (edible and choice), *Amanita citrine* (probably edible), and the *Aminita verna* (more commonly known as the Destroying Angel).

We begin by storing the relative losses in a matrix with zeros along its diagonal:

```
> M=c(0,1, 2,100,0, 50,1,2,0)
> dim(M)=c(3,3
> rpart(species~CapLth + Volvath + Capwth
  + Volvasiz, parms=list(loss=M)))
 1) root 150 150 A.verna
    (0.3333333 0.3333333 0.3333333)
 2) Volvasiz< 0.8 50 0 A.citrina
    (1.0000000 0.0000000 0.0000000) *
 3) Volvasiz>=0.8 100 100 A.verna
    (0.0000000 0.5000000 0.5000000)
 6) Volvasiz< 1.85 66 32 A.verna
    (0.0000000 0.7575758 0.2424242)
12) Volvath< 5.3 58 16 A.verna
    (0.0000000 0.8620690 0.1379310) *
13) Volvath>=5.3 8 0 A.calyptroderma
    (0.0000000 0.0000000 1.0000000) *
 7) Volvasiz>=1.85 34 0 A.calyptroderma
    (0.0000000 0.0000000 1.0000000) *
```

Note that the only assignments to edible species (the group of three values in parentheses at the end of each line) made in this decision tree are those which were unambiguous in the original sample. All instances, which by virtue of their measurements, might be confused with the extremely poisonous *Amanita verna* are classified as *Amanita verna*.

Minimizing the Cost of Data Collection

While a research grant may have permitted us the luxury of gathering a large amount of data on a limited number of subjects when we formulated our original model, we may not have quite the same luxury when it comes to applying it.

Take a second look at the Boston data, and let us assume that we'd like to apply it to our own city today. The data for certain of the predictors can be readily gathered from county records, such as CRIM and TAX. Measuring

NOX would be extremely time consuming. Fortunately, NOX does not appear in the model. Measuring DIS and RAD would require just a little extra effort. As for RM, the average number of rooms per dwelling, collating modification requests with the original building permit would also be quite time consuming (assuming, that is, that rooms were never added on without a permit).

To create a model that takes these relative costs into account:

```
> data=read.table("http://statcourse.com/
  research/boston.csv", , sep=",", header
  = TRUE)
> library(rpart)
> fit=rpart (MEDV~CRIM+ZN+INDUS+CHAS+NOX+
  RM+AGE+DIS+RAD+TAX+ PT+B+LSTAT,
  cost = c(1, 2 , 2, 1, 10, 4, 1, 2.2, 2,
  2, 1, 2.5, 1))
> print(fit)
Resulting in a quite different tree
(Figure 14.4):
  1) root 506 42716.3000 22.53281
  2) LSTAT>=9.725 294 7006.2830 17.34354
  4) LSTAT>=16.085 144 2699.2200 14.26181
  8) CRIM>=5.77 74 1043.6960 11.97568 *
  9) CRIM< 5.77 70 859.9179 16.67857 *
  5) LSTAT< 16.085 150 1626.6090 20.30200 *
  3) LSTAT< 9.725 212 16813.8200 29.72925
  6) LSTAT>=4.65 162 6924.4230 26.64630
 12) PT>=13.85 152 4642.1460 25.81250
 24) AGE< 89.45 144 3240.5380 25.37917
 48) LSTAT>=7.685 57 914.4484 23.00526 *
 49) LSTAT< 7.685 87 1794.4170 26.93448
 98) PT>=18.85 28 109.2211 24.01786 *
 99) PT< 18.85 59 1333.9690 28.31864
198) RM< 6.775 34 197.0894 25.18235 *
199) RM>=6.775 25 347.6136 32.58400 *
 25) AGE>=89.45 8 887.8487 33.61250 *
 13) PT< 13.85 10 570.3760 39.32000 *
  7) LSTAT< 4.65 50 3360.8940 39.71800
```

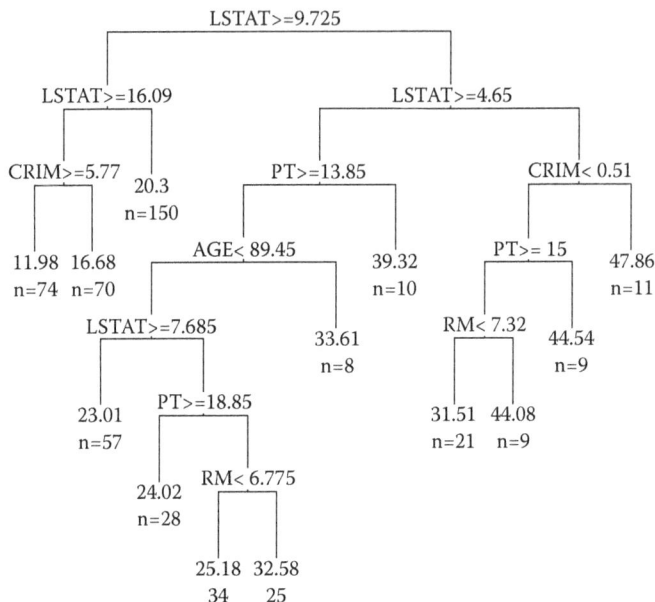

FIGURE 14.4 Decision tree for the Boston data after correcting for the relative costs of data collection.

```
14)  CRIM< 0.51 39 2288.5440 37.42051
28)  PT>=15 30 1371.8820 35.28333
56)  RM< 7.32 21 244.5857 31.51429 *
57)  RM>=7.32 9 132.8956 44.07778 *
29)  PT< 15 9 322.8822 44.54444 *
```

Using the Decision Tree as an Aid to Decision Making

Suppose we have gathered the data for another house to which we want to apply the previous decision tree. To avoid problems with R, we use the existing data set as a template:

```
> new=edit(data[1,])
> new
```

```
     CRIM ZN INDUS CHAS NOX RM  AGE  DIS RAD TAX PTRATIO B     LSTAT MEDV
1 0.006 NA NA     0    NA  7.0 65.2 NA  NA  NA  14      396.9 10.5  NA
2 0.030 0  2.18   1    NA  6.4 46.0 NA  NA  222 18      NA    3.0   NA
```

Note that either we entered an NA for not available whenever we didn't collect a data item or merely left the existing entry untouched when it would not be used in the decision tree.

```
> predict(fit,new)
     1      2
20.302 31.514
```

Summary

In this chapter, you were shown how to construct and apply decision trees to both classification and regression. In both instances, it was possible to modify the resultant tree to correct for the costs of data collection. Classification trees can be modified to provide for the prior probabilities of the various classes and the relative costs of misclassification.

To Learn More

See Good (2011a) for more on the application of decision trees to the analysis of microarrays and photographic imagery. See Good (2011b) for alternatives to the tree-construction method described in this chapter.

Comparisons of the regression and tree approaches were made by Nurminen (2003) and Perlich et al. (2003).

PART V

REPORTING YOUR RESULTS

CHAPTER **15**

Reports

Prescription

1. Choose a journal to which to submit your report.
2. Consult an article or two from this journal, and use these articles as a template. Typically, you'll be asked to provide an Abstract, Introduction, Methods and Materials, Results, Discussion, and References.
3. Defer working on the Abstract and Introduction sections until the rest of your report is complete. Start with the Methods and Materials section.
4. Prepare an outline of your Results. Decide for each finding whether it is best conveyed by text, table, or graph.
5. Prepare a brief Discussion of your findings, noting how they may support, advance, or contradict previous work.
6. Prepare the Introduction.
7. Write the Abstract.
8. Consult the Guidelines of the journal to which you hope to submit your report, and make sure the citations in your Bibliography are in accordance with those guidelines.

Choose a Journal

Consult a database of journal impact factors (such as the Web of Knowledge) and sort the journals for the appropriate subject area in descending order. Select an optimistically high impact factor journal to start with, then, depending on the response received, work down the list from there.

Depending on how quickly a manuscript needs to be published and how complete/novel the work is, it may be worth considering submitting it as a note or short communication, rather than as a full article. These different forms of manuscript will have different limits on the number of tables/figures/references, so this should be considered when drafting the manuscript.

Methods and Materials

Your methods and materials need be described in sufficient detail that they can be readily reproduced in other investigators' laboratories.

Robert Boyle, 1661

Provide all of the following:

- Quality control measures
- Details of power and sample size calculations
- The experimental design, the number of subjects per treatment, treatments administered, choice of controls, and treatment intervals
- Details of treatment allocation
- Type and adequacy of blinding

Use short sentences, and try to keep your descriptions jargon free. Hopefully, you've followed the advice we gave in Chapter 5, and your methods and materials section is practically complete already.

Treatment Allocation*

Treatment allocation must be described in full:

Was allocation discretion available to the investigator, the subjects, both, or neither (as in dictated allocation)? Were investigators permitted to assign treatment based on subject characteristics? Could subjects select their own treatment from among a given set of choices?

Was actual (not virtual, quasi-, or pseudo-) randomization employed? Was the allocation sequence predictable? Why not?

Was randomization *conventional,* that is, the allocation sequence was generated in advance of screening subjects? Why not?

Was allocation concealed prior to its being executed?

Was randomization unrestricted in that a subject's likelihood of receiving a specific treatment was independent of all previous allocations or restricted?

Were treatment codes concealed until all patients had been randomized and the database locked? Were there instances of codes being revealed accidentally?

Was the allocation sequence ever violated, deliberately or accidentally?

Was randomization simultaneous, block simultaneous, or *sequential?*

Was intent to treat permitted, that is, was treatment ever modified after allocation?

Blinding

Describe whether blinding was single, double, or triple; how blinding was enforced; and any violations.

---

* The material in this section relies heavily on a personal communication from Vance W. Berger and Costas A. Christophi.

Treatments

Describe treatments in sufficient detail that they are readily replicated. Describe how they were administered, the nature of the experimental design, and whether and how many treatments were missed, incomplete, or switched.

Surveys

Describe how surveys were administered and how they were monitored. Provide links to sample forms. Describe any and all follow-ups that were made to nonresponders and for verification purposes

Results

Descriptive Statistics

In this section, we consider how to most effectively summarize your data whether it comprises a sample or the entire population.

Binomial Trials

The most effective way of summarizing the results of a series of binomial trials is by recording the number of trials and the number of successes, for example, the number of coin flips, the number of heads, the number of patients treated, the number who got better, and so forth. Percentages can be misleading.

When trials can have three to five possible outcomes (multinomial), the results are best presented in tabular form (as in Table 15.1) or in the form of a bar chart (see Figure 17.1), whether the outcomes are ordered (no effect, small effect, large effect) or unordered (win, lose, tie). Both forms also provide for side-by-side comparisons of several sets of trials.

For the reasons discussed in the next chapter, we do *not* recommend the use of pie charts.

TABLE 15.1   RBI's per Game

	0	1	2	>2	Didn't play
Good	2	3	2	1	3
Hardin	1	2	1	4	0

TABLE 15.2   Sandoz Drug Data

	New drug		Control drug	
Test Site	Response	#	Response	#
1	0	15	0	15
2	0	39	6	32
3	1	20	3	18
4	1	14	2	15
5	1	20	2	19
6	0	12	2	10
7	3	49	10	42
8	0	19	2	17
9	1	14	0	15

Categorical Data

Display the results for a single categorical variable in the form of a bar chart. If there are multiple variables to be considered, the best way to display the results is in the form of a contingency table as shown in Table 15.2.

We can also summarize a single 2 × 2 table, simply by reporting the odds ratio

$$\frac{p(1-s)}{(1-p)s}$$

where p is the proportion of successes in one of the categories and s is the proportion of successes in the remaining category. If the categories are independent, then the odds ratio will be unity.

Rare Events

Reporting on events that are rare and random in time and/ or space like suicides, drownings, radioactive decay, the seeding of trees and thistles by the winds, and the sales of my novels* can be done in any of three different ways:

1. A statement as to the average interval between events—a day in the case of my novels.
2. A statement as to the average number of events per interval—50 per month in the case of my novels.
3. A listing in contingency table form of the frequency distribution of the events (see Table 15.1).

The clustering of random events is to be expected and *not* to be remarked upon. As a concrete example, while the physical laws which govern the universe are thought to be everywhere the same, the distribution of stars and galaxies is far from uniform; stars and galaxies are to be found everywhere in clusters and clusters of clusters (see Neyman and Scott, 1952).

Measurements

To quickly summarize a group of measurements provide a measure of their central tendency and a measure of their dispersion.

Measures of central tendency include the arithmetic mean, the geometric mean, or the median.

For small samples of 3–5 observations, summary statistics are virtually meaningless. Reproduce the actual observations; this is easier to do and more informative.

Consider reporting the median rather than the arithmetic mean or average as the latter can be very misleading. For example, the mean income in most countries is far in excess of the *median* income or 50th percentile to

---

* Search for *Sad and Angry Man* or Kindle Books *Luke Jackson* at http://www.amazon. com or http://www.zanybooks.com.

which most of us can relate. When the arithmetic mean is meaningful, it is usually equal to or close to the median.

The purpose of your inquiry must be kept in mind. The distribution of orders in dollars from a machinery plant is likely to be skewed by a few large orders. The median dollar value will be of interest in describing sales and appraising salespeople; the mean dollar value will be of interest in estimating revenues and profits.

The *geometric mean* is more appropriate than the arithmetic in three sets of circumstances:

1. When losses or gains can best be expressed as a percentage rather than a fixed value.
2. When rapid growth is involved as is the case with bacterial and viral populations.
3. When the data span several orders of magnitude, as with the concentration of pollutants.

Whatever statistic you use, be sure to report only a sensible number of decimal places. If your observations were to the nearest integer, your report on the mean should include only a single decimal place.

Most real-life populations are actually mixtures of several homogeneous populations. If multiple modes are observed in samples greater than 25 in size, the number of modes should be reported. A plot of the frequency distribution would also help the reader to interpret your findings.

Measures of dispersion include the variance, the mean absolute deviation, the inter-quartile range, and the range. The standard error is a useful measure of population dispersion *if* the observations are continuous measurements that come from a normal or Gaussian distribution. If the observations are normally distributed, then in 95% of the samples we would expect the sample mean to lie within two standard errors of the mean of our original sample.

But if the observations come from a nonsymmetric distribution like an exponential or a *Chi*-square, or a truncated distribution like the uniform, or a mixture of populations, we cannot draw any such inference.

Recall that the standard error equals the standard deviation divided by the square root of the sample size, SD/√n or

As the standard error depends on the squares of individual observations, it is particularly sensitive to outliers. A few large observations will have a dramatic impact on its value.

A loose rule of thumb is that the mean of a sample of 8 to 25 observations will have a distribution that is close enough to a normal distribution for the standard error to be meaningful. The more nonsymmetric the original distribution, the larger the sample size required. At least 25 observations are needed for a binomial distribution with $p = 0.1$.

Even the mean of observations taken from a mixture of distributions (males and females, tall Zulu and short Bantu)—visualize a distribution curve resembling a camel with multiple humps—will have a normal distribution if the sample size is large enough. Of course, this mean (or even the median) conceals the fact that the sample was taken from a mixture of distributions.

If the underlying distribution is not symmetric, the use of the ±SE notation can be deceptive as it suggests a nonexistent symmetry. For samples from nonsymmetric distributions of size 6 or less, tabulate the minimum, the median, and the maximum. For samples of size 7 and up, consider using a box and whiskers plot. For samples of size 16 and up, bootstrap estimates of the percentiles may provide the answer you need.

Report the numbers of truncated observations as well as the truncated values. Describe the methods of imputation of missing data if any were employed.

Ordinal Data

For ordinal data, an arithmetic average or a variance would not be at all meaningful. One can report such results in tabular form, in bar charts, or by providing key percentiles such as the minimum, median, and maximum. With questionnaires where there are only 3 to 5 alternatives per question, bar charts are recommended.

Survival and Mean-Time-to-Failure Data

Plot the results. Report the number of censored observations. Report the model that was employed, the hazard function, and the median or mean survival times.

Missing Data

Every experiment or survey has its exceptions. You must report the raw numbers of such exceptions and provide additional analyses that analyze or compensate for them. Typical exceptions and potential remedies are listed below:

1. Subjects who were eligible and available but did not participate in the study. With mail-in surveys, distinguish between those whose envelopes were returned "address unknown" and those who simply did not reply. In each instance, a follow-up survey of the nonresponders is called for.

   The definition of "nonresponder" also applies to partially completed questionnaires where the answers to key questions were omitted.

2. Subjects who were randomized to treatment, but later determined to lie outside the study population. Present the final analysis in two parts, one incorporating all patients, the other limited to those who were actually eligible.

3. Subjects who enrolled in the study but did not complete it including both dropouts and noncompliant patients. These subjects might be subdivided further based on the point in the study at which they dropped out or went missing. Traditional statistical methods are not applicable when withdrawals are treatment related.

4. If the design provided for intent to treat and noncompliant subjects were continued in the study after

being assigned to an alternate treatment, two sets of results should be reported: the first for all patients who completed the trials (retaining their original treatment assignments for the purpose of analysis); the second restricted to the smaller number patients who persisted in the treatment groups to which they were originally assigned.

5. Censored and off-scale measurements should be described separately and their numbers indicated in the corresponding tables.

You need to analyze the data to ensure that the proportions of missing observations are the same in all treatment groups. Again, traditional statistical methods are applicable only if missing data are not treatment related.

Tables

Is text, a graph, or a table the best means of presenting results? Dyke (1997) would argue, "Tables with appropriate marginal means are often the best method of presenting results, occasionally replaced (or supplemented) by diagrams, usually graphs or histograms." van Belle (2002) warns that aberrant values often can be more apparent in graphical form.

Not all data need be tabulated in your report, but the raw data should be made available via the Internet. The program listings for simulations should also be made available.

Text should be used for displaying two to five numbers, as in "The blood type of the population of the United States is approximately 45% O, 40% A, 11% B, and 4% AB."

Graphs should be used only if they are indeed worth a thousand words. Never use a chart that will take longer to explain than the information it was intended to provide.

"Use the table heading to convey critical information. Do not stint. The more informative the heading, the better the table."[*] Sample sizes should always be specified.

---

[*] vanBelle (2002, p. 154).

Confidence limits and standard errors should be clearly associated with the correct set of figures.

Results should be expressed in the appropriate units, and rounded off to the correct degree of precision.

Residuals should be tabulated; the resulting table can alert us to the presence of outliers and may also reveal patterns in the data.

Limit tables to one or two factors. Tables involving three or more factors are not always immediately clear to the reader and are best avoided.

Reporting Your Analyses

Describe the statistical tests you employed along with the assumptions inherent in your choices. Whether you report p-values or confidence intervals, be cautious in your interpretations.

p-Values? or Confidence Intervals?

"The p-value is *not* the probability that the null hypothesis is true" (Yoccuz 1991).

The p-value is a random variable that varies from sample to sample. Two populations may have many differences of practical significance, and yet the samples taken from those populations and the resulting p-value may not reveal that difference. Consequently, it is not appropriate to compare the p-values from two distinct experiments, or from tests on two variables measured in the same experiment, and declare that one is more significant than the other.

If we agree in advance of examining the data that we will reject the hypothesis if the p value is less than 5%, then our significance level is 5%. Whether our p-value proves to be 4.9% or 1% or 0.001%, we will come to the same conclusion. One set of results is not more significant than another; it is only that the difference we uncovered was measurably more extreme in one set of samples than in another.

TABLE 15.3 *p*-value versus association

p-value	Gamma		
	<0.30	**0.30 to 0.70**	**>0.70**
<0.1	8	11	5
0.05	7	0	0
>0.10	8	0	0

Source: Data, T. J. and Dean, C. W., *Amer. Sociologist.*, 45–46, 1968, February.

p-values need not reflect the strength of a relationship. Duggan and Dean (1968) reviewed 45 articles that had appeared in sociology journals between 1955 and 1965 in which the *Chi*-square statistic and distribution had been employed in the analysis of 3 × 3 contingency tables and compared the resulting *p*-values with association as measured by Goodman and Kruskal's gamma. Table 15.3 summarizes their findings.

The vast majority of *p*-values produced by parametric tests are approximations. A stated significance level of 4.9% might really prove to be 5.1% in practice.

Regardless of which test one uses, it is the height of foolishness to report *p*-values with excessive precision; 0.06 and 0.052 are both acceptable, but 0.05312 suggests you've let your software do the thinking for you.

If *p*-values are misleading, what are we to use in their place? Jones (1955, p. 407) was among the first to suggest that "an investigator would be misled less frequently and would be more likely to obtain the information he seeks were he to formulate his experimental problems in terms of the estimation of population parameters, with the establishment of confidence intervals about the estimated values, rather than in terms of a null hypothesis against all possible alternatives."

Confidence intervals can be used both to evaluate and report on the precision of estimates and the significance of hypothesis tests. The probability the interval covers the true value of the parameter of interest and the method used to derive the interval must also be reported.

On the other hand, like the *p*-value, the upper and lower confidence limits of a particular confidence interval are random variables for they depend upon the sample that is drawn. The center of the interval is no more likely than any other value, and the confidence to be placed in the interval is no greater than the confidence we have in the experimental design and statistical test it is based upon.

Multiple Tests

Whether we report *p*-values or confidence intervals, we need to correct for multiple tests as described in Chapter 9. The correction should be based on the number of tests we *perform*, which in most cases will be larger than the number on which we report.

Discussion

Comment on your findings in relation to those of previous investigators, and discuss the implications for future research.

Discuss errors in your methodology, for example, treatment allocation codes being revealed accidentally, as well as how such accidents might be avoided.

Discuss possible sources of bias.

Introduction

Your introduction should contain a brief summary, no more than a paragraph per point, of what you hoped to learn and why, of the methods you planned to use, and what you found out. In contrast to the introduction to a dissertation, term paper, or review article, it does *not* contain a comprehensive history of all previous work on the subject. It should include citations to previously published work that you hope to confirm or refute.

Abstract

The abstract of your article should contain a brief summary of your methods and your findings. A second paragraph is rarely justified.

Bibliography

As someone who has labored long and hard over the preparation of bibliographies (and then had to do the work all over again in order to submit the work to a different journal), I am deeply grateful to Tony Rowe for pointing out that a bibliography program such as EndNote and Reference Manager is as essential to the working researcher as a word processor or a spreadsheet.

Responding to Rejection

While you no doubt assume that the return post will bring a letter from the journal editor telling you that your article has not only been accepted but a special edition of the journal will be devoted to it, the more likely possibility is that apart from acknowledgment of receipt, you'll hear nothing for three months. After that time, one ought contact the editor (by mail or e-mail) and inquire as to the present status of one's article. The editor's reply will boil down to "I'll look into it."

Subsequently, you may receive not the letter of acceptance you'd hoped for, but one of conditional acceptance or outright rejection. Conditional acceptance will usually require that you make some changes. In most instances, you will agree with reviewers' suggestions. If some of their concerns strike you as off the wall, could it be that your writing lacks clarity and reviewers are misled? A substantial rewrite may be called for.

If you disagree with a proposed change, and can argue your case in one or two sentences, then do so in a reply to the editor along with a copy of your article that you have revised in accordance with the balance of the proposed modifications.

If you disagree with more than one of the proposals, consult several of your colleagues before proceeding further. It may be that additional supporting studies are called for. This same advice applies if you receive a letter of outright rejection. Read and reread the reviewer's comments. If in fact, the results of your study are open to more than one interpretation, then you may need to do an additional study before submitting your findings for publication.

One the other hand, the reviewer simply may be a total ass. I once received a review which began, "Although I have not read this paper, I judge from the title that it would not be of particular interest to this journal."

The appropriate response to such a review is to submit your article to another journal, after first reading that new journal's submission guidelines and reformatting your article (and, in particular, its bibliography) to meet them.

To Learn More

For more on reporting requirements, see Bailar and Mosteller (1988), Begg et al. (1996), Grant (1989), Altman et al. (2001), http://www.aera.net/uploaded-Files/Opportunities/StandardsforReportingEmpirical SocialScience_PDF.pdf, and International Committee of Medical Journal Editors (1997). Reporting criteria for meta-analyses are given in Lang and Secic (1997; p177ff).

For guides to the appropriate number of digits to use in your reports, see Ehrenberg (1977) and van Belle (2002; Table 7.4). For advice on getting your manuscript accepted, see Boyd et al. (2007).

On the proper role of p-values, see Neyman (1977) and Cox (1977).

See Tufte (1983) on the issue of table versus graph.

Good (2009; Chapter 8) provides an extensive list of potential sources of bias.

CHAPTER 16

Oral Presentations

Prescription

Build your presentation around a series of slides.
Restrict each slide either to three key phrases, two photos, two graphs, or single table.

Plan on presenting a series of slides; show a slide first and then discuss it. The first slide should contain the lecture title, the name of the speaker, and sufficient information (an e-mail address, for example), that the speaker may be contacted for further information. The second slide should contain a brief summary of what is to be discussed. If a series of lectures is contemplated, this slide will summarize the series, and would be followed by a summary slide of the first lecture. The next-to-final slide is a repeat or variant of this summary, and the final slide is a repeat or variant of the title slide.

Text

Each slide requires a title and no more than three bullet points in the body of the slide. If a fourth summary bullet point is required, make use of animation features to have

Power Depends On

- Sample Size
- Alternative
  - Size of Effect
  - Underlying Variation
- Significance Level

FIGURE 16.1

this final point slide or drop in after the balance of the slide had been discussed.

Each bullet point should be restricted to a single key phrase as in Figure 16.1. Do not rehash or recopy the lecture itself.

Graphs

The rules for graphics set forth in the next chapter apply equally to lectures. In brief, "Never use a chart that will take longer to explain than the information it was intended to provide."

Use two graphs or two photos on the same slide only if a side-by-side comparison is intended.

If colors are used, in a pie chart, for example, check to be sure they can be readily distinguished when they appear in shades of gray on handouts.

Tables

Restrict the table's title to two lines of text that completely identify the table's contents.

The numeric values in a table should occupy no more than three or four columns and be limited no more than three digits each, for example, 318, 3.18, 3.1×10^8.

No footnotes. If items in the table need to be starred, plan to explain the stars when you display the table.

Chapter 17

Better Graphics

Rows and flows of lines and spots,
Histograms and digidots,
Box-and-whiskers, quantile plots
I've looked at stats that way
But is it just a picture show?
If something's there how will you know?
And if they say it isn't so
What is there left to say?
I've looked at plots from both sides now,
Transformed and spun them: still, somehow,
It's plots' illusions I recall,
I really can't trust plots at all.

Robert Dawson, writing in LabLit

Prescription

KISS: A picture is easily worth a 1000 words, but not if it will take more than 1000 words to explain its purpose. Keep your graphics simple, and ensure that each graph is sufficiently complete that it can stand alone if separated from the surrounding text.

To achieve this goal, do all of the following:

- Use the same number of dimensions as in the information to be illustrated. Use three dimensions only if trying to depict the values of three variables simultaneously. Even then, could the same objectives be accomplished with contours in preference to perspective or by labeling the data points with the values of the third variable?
- Use color sparingly. In fact, keep the number of line styles, shades, and symbols to a minimum. (Still, color can revive a half-asleep audience at the end of a long lecture.)
- Match the range of the axes to the values to be displayed. Exception: When relative as well as absolute comparisons should be made by the reader.
- Only label the axis with ruled measurements when the data values are also continuous.
- Connect data points only when interpolation would be meaningful.
- Place labels and legends in otherwise unused portions of the plotting region.
- Order the symbols in the legend in the same order they appear in the graphic.
- A figure's caption must include the who, what, when, and where of its data's origins. It's all too easy for a figure to become separated from the main body of the text.
- Avoid jargon. Use plain English in descriptions as well as in the body of your article.

Creating Graphs with R

Though R has more than 20 packages (each providing multiple functions) that one can download and use to create graphs and charts to enhance your reports, in light of our repeated advice to keep things simple, we focus in this section on just four functions: `barplot()`

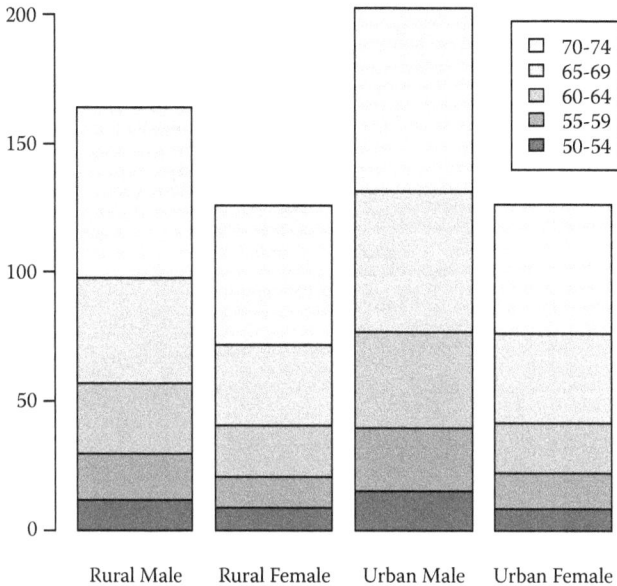

FIGURE 17.1 Bar chart of Veterans Administration deaths by sex.

for categorical data, and `boxplot()`, `dotplot()`, and `plot()` for measurements. As we shall soon see, some come with helper functions that facilitate communication in line with the preceding guidelines.

VADeaths, the data used to construct the barplot depicted in Figure 17.1, is part of the standard download of R.

```
> barplot(VADeaths, col=gray.colors(5),
  legend = rownames(VADeaths))
```

The box and whiskers plot in Figure 7.2 was created with the command

```
> boxplot(classdata)
```

A stripchart can be overlaid as in Figure 7.4 with the command

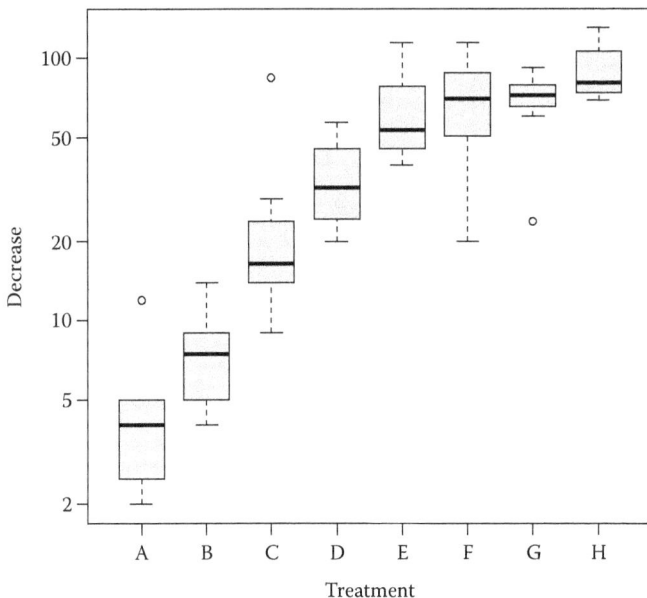

FIGURE 17.2 Effect of treatment on the decrease of orchard infestations.

```
> stripchart(classdata)
```

The side-by-side box and whiskers plots shown in Figure 17.2 were created with

```
> boxplot(decrease~treatment,
  data=OrchardSprays, log="y",col="light
  gray",boxwex=0.7, xlab="treatment",
  ylab="decrease")
```

where the dataframe OrchardSprays is included when one downloads R, decrease and treatment are two of the variables in OrchardSprays, log = "y" specifies that decrease will be plotted on a logarithmic scale along the *y*-axis, boxwex = 0.7 reduces the width of the individual box plots and increases the separation between them, and xlab and ylab provide labels for the axis.

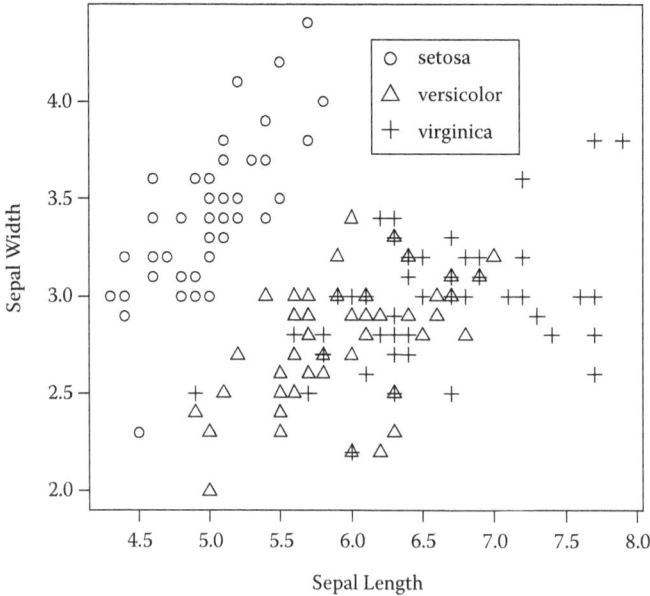

FIGURE 17.3   Scatter plot helps classify Iris into distinct species.

Scatter plots have multiple uses. In Figure 17.3, a scatter plot helps classify Iris into distinct species.

```
#use the Iris data set
> with(iris,
+ plot(Sepal.Length, Sepal.Width,pch=as.
numeric(Species), cex=1.2))
#as.numeric (Species): assign the
  numerals 1,2,3 to the Species in their
  order of appearance.
#pch: plot points using one of the symbols
  1,2, or 3
#cex: enlarge the symbols by the factor
  of 1.2
> legend(6.1,4.4,c("setosa","versicolor",
  "virginica"), cex=1.5,pch=1:3)
#place the legend out of the way, with the
  upper left hand corner at x=6.1, and y=4.4
```

Figure 17.4a looks ordinary enough, except that the numerous duplicate values are hidden from view. To bring these duplicates to light, we need only add a proportionally small random value to each observation. Figure 17.4b was created with the command plot(jitter(Xr),jitter(Yr)).

The cumulative distribution of the normal is plotted in Figure 7.2 with the command

```
> plot(z,y,type="l",main="Cumulative
  Distribution",ylab="percentile")
```

The empirical distribution function of a sample taken from a normal distribution is plotted in Figures 17.5a,b using the following commands:

```
> plot(y,ppoints(y),type ="s", xlab="x",
  ylab="F[x]")
> plot(sort(y),ppoints(y),type ="s",
  xlab="x", ylab="F[x]")
```

As can be seen in these figures, the points need to be sorted if we are to connect them by straight lines.

In Figure 17.6, which compares the cumulative distribution functions of successively larger samples, we overlay three curves on a single scatter plot as follows:

```
#Generate three successively larger
 samples from a Normal distribution
> y=rnorm(10); w=rnorm(100); z=rnorm(1000)
#Plot the cumulative distribution of the
 mid-size sample
> plot(sort(w),ppoints(w),type ="s",
 xlab="x", ylab="F[x]")
#Add a title
> mtext("Empirical Distributions of
 Samples from a Normal Distribution",3)
#Plot points on the graph corresponding to
 the other two samples
> points(sort(y),ppoints(y),pch =3)
```

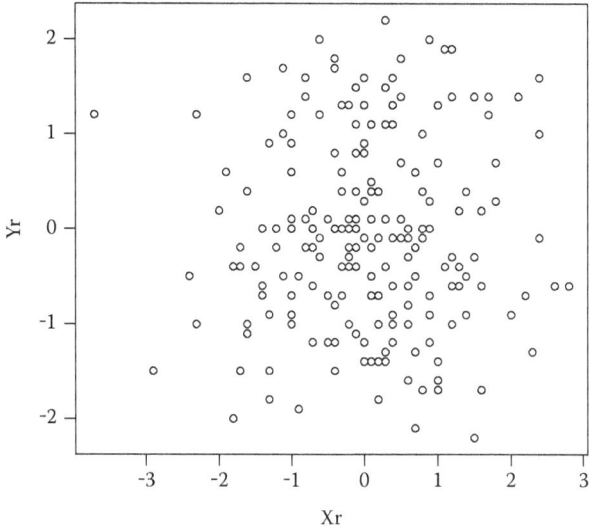

(a)

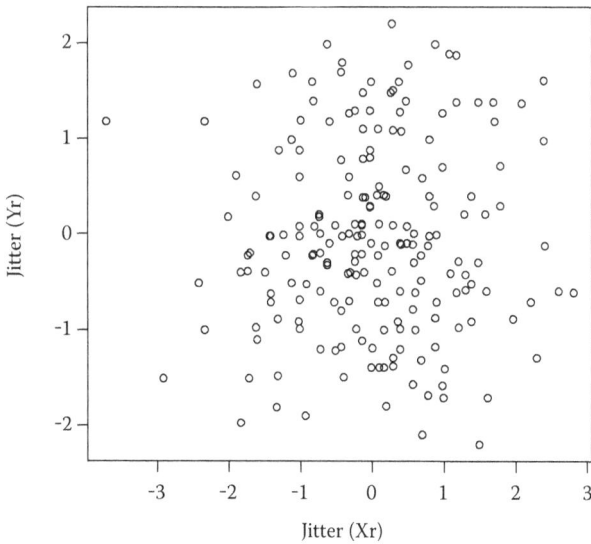

(b)

FIGURE 17.4 Using R's `jitter()` command to reveal duplicate values.

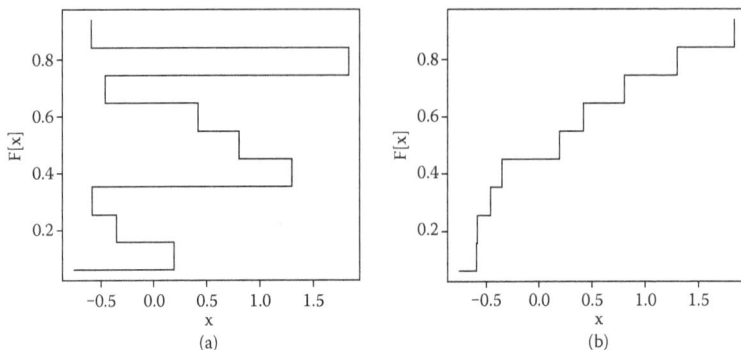

FIGURE 17.5 (a) Plotting unsorted data; (b) Plotting sorted data.

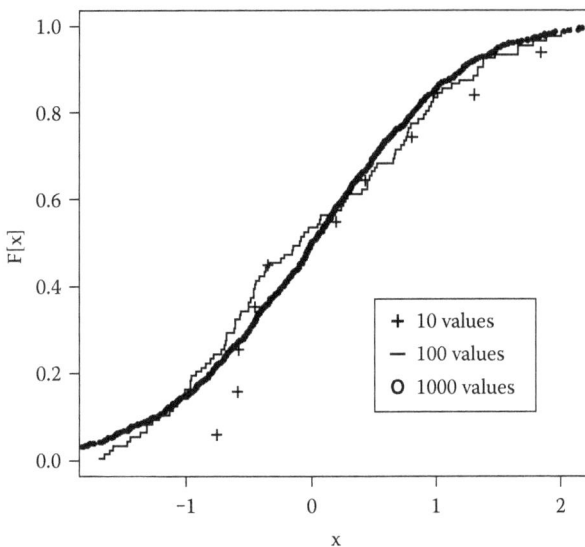

FIGURE 17.6 The distributions of successively larger samples get closer and closer to the distribution of the population from which they were drawn.

```
> points(sort(z),ppoints(z),pch =19,
cex=0.5)
#Add a legend
> legend(0.5,0.4,c("10 values",
"100 values","1000 values"), cex=1.5,
pch=c("+","_","o"))
```

As Figure 12.7 demonstrates, a third dimension can be added to a scatter plot by making use of the `symbols()` function instead of plot:

```
symbols(Age, QUI, circles=CHOL, inches=F)
```

To Learn More

To explore the full range of graphic possibilities using R, consult Murrell (2011). For displaying questionnaire data, see Falissard (2012). Through numerous examples, Cleveland (1994) and Wilkinson (1999) will help you to make the most effective use of each graphic element. Mosteller and Tukey (1977) as well as Hardin and Hilbe (2007, pp. 143–167) use graphics for assessing model accuracy.

PART VI
NONRANDOM SAMPLES

Chapter 18

Cohort and Case-Control Studies

Case-control and cohort studies can be invaluable in assessing the effects of government intervention or of the necessity for such intervention. In these studies, we compare outcomes in two cohorts (samples) drawn from different time periods or different regions with the aim of attributing the differences in the outcomes to the differences in the environments.

They are considered here in a separate chapter because they differ in all aspects from randomized studies. They differ in how samples ought to be collected—for we cannot assign the treatments at random, to how they ought to be analyzed, and to how the results ought to be reported. Also, a slightly different vocabulary is often employed.

A Worked-Through Example

Suppose we wish to know how coffee drinkers differ from non-coffee drinkers. Assigning subjects at random to drink or not-drink coffee is unlikely to succeed. We begin by conducting a survey without revealing its purpose and compare the characteristics of one group with that of another. This comparison might involve side-by-side box or pie charts or side-by-side tables. Our report would include not only the estimates for each sample, but confidence intervals for these estimates. For 2 × 2 tables, report the odds ratio (the relative risk). Do not compute *p*-values, because treatments were not assigned to the cohorts at random.

We can increase the efficiency of such comparisons by pairing each coffee drinker with a non-coffee drinker on several baseline attributes such as sex, age, and current health, thus reducing the sources of variation as we would in any other matched-pair study. On the downside, we might have fewer subjects to observe with such a *case-control* study.

We can and perhaps should continue to follow the two-groups of subjects so as to compare the relative health of their members at a future date. Anticipate dropouts and incorporate plans for retention at the design stage.

Prescription

- Formulate hypotheses with one data set; analyze with another. (If necessary, split the original sample in half at random before examining either half in detail).
- Take care in selecting populations (frames) from which to draw samples to ensure they are relatively homogeneous and will generalize to other populations or subpopulations.
- Choose a control cohort that differs in as few attributes as possible from the cohort of interest.
- To locate subjects with a specific property (for example, coffee drinkers), ask for volunteers to take a survey before revealing the purpose of the survey.
- Report all unanticipated differences between cohorts.
- When developing models, avoid overmatching through the use of extraneous variables that are both risk factors for the effect being studied, and associated with the exposure being studied though not a consequence of exposure.

■ When making retrospective comparisons, be wary of *recall bias*, the propensity of diseased subjects when interviewed to scrutinize their memory and report more accurately on past exposure and possible causes of their disease than nondiseased subjects would do.

Examples

Smith and Douglas (1986) analyzed the incidence of leukemia of the cohort of workers at a British Nuclear Fuels plant to examine the effects of occupational exposure to radiation. The "treatment" cohort consisted of workers handling radioactive material. The control cohort consisted of office workers employed at the same plant during the same period. Case matching was based on sex and date of birth (within two years). Morbidity was found to be correlated with occupational status. But when case matching was also made with an individual's initial date of employment, the correlation disappeared as the result of overmatching, for radiation dose also changes with calendar time.

Marshall et al. (2011) examined the population-based overdose mortality rates in two distinct areas of Vancouver for the three-year periods before and after the opening of the Vancouver Safe-Injection Facility. They reported a practical as well as statistically significant decrease in the immediate (500-meter) area in contrast to a minor decrease in the fatal overdose rate in the rest of the city.

Their choice of control period is suspect; one of the control years had markedly higher heroin availability and overdose fatalities than all subsequent years. Other changes in government policy may have affected the results; 50–66 extra police were specifically assigned to the 12 city blocks surrounding the safe-injection facility during the after period.

To Learn More

See Parfrey and Barrett (2009) and Thomas (2009).

R Primer

If you've never used a programming language before, don't panic. This primer will enable you to make immediate use of all the R routines in the balance of this text. If you feel you require additional assistance, I teach an Introduction to R course online during which I'm available on a daily basis to answer questions. To enroll, access http://statcourse.com/intro2R.htm.

Begin by downloading the R package without charge from the website http://cran.r-project.org/. You will need only the binary files specific to your computer's operating system (Windows, Linux, or Macintosh).

R is an *interpreter*. This means as you enter the lines of a typical program, you learn line by line whether the command you've just entered makes sense (to the computer) and are able to correct the line if you have made a typing error. (The up arrow on the PC ^ will restore the previous line.)

Run R. What you see at the bottom of the screen is an arrowhead (>).

Type 2 + 3 after the arrowhead and then press the Enter key:

[1] 5.

R reports numeric results in the form of a vector. In this example, the first and only element in the answer vector [1] takes the value 5. If you'd wanted to save the result for later use, you might have written

result = 2 + 3

Exercise : What is (2+3)/(233-219)*7?

R permits you to add two vectors with a single command:

```
> x=1:5
> y=c(2,4,5.1,3,1)
> x+y
[1] 3.0 6.0 8.1 5.0 8.0
```

R can't read your mind, nor will it correct your spelling. If you type X instead of x,

```
> X+y
Error: object 'X' not found
```

R comes with a large number of built-in functions. You've already met the concatenate function, c( ). Midway through the text, you'll encounter mean(x) and sd(x).

Many R functions you'll encounter in this text have a large number of capabilities that this text may not make full use of. To make use of these "hidden" capabilities, pull down R's Help menu, and select and click on R functions (text). Type in the name of a function, and you'll get a full description of its capabilities along with examples of its use.

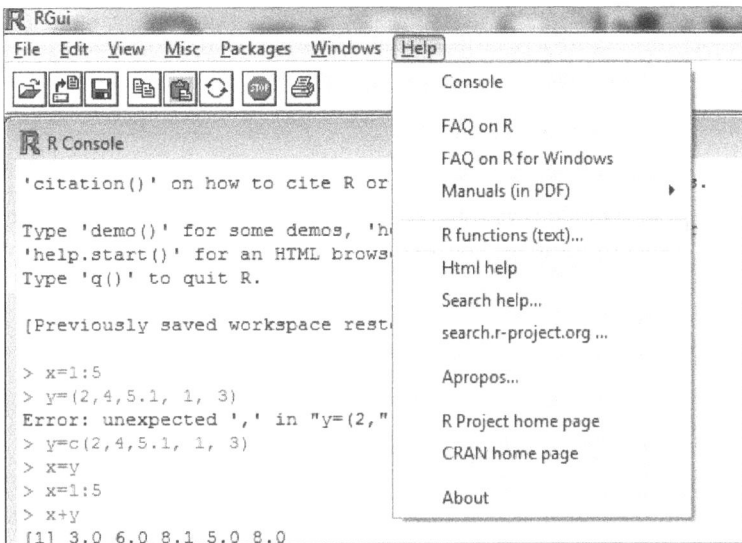

In addition to the basic sets of functions that come with R, this text makes use of many packages of special functions that can be downloaded from the Internet by selecting the packages menu and clicking on Install packages.

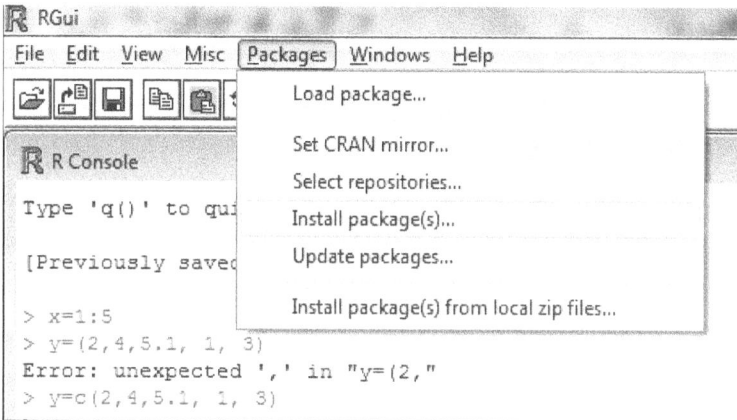

Bibliography

Altman, D. G., Schulz, K. F., Moher, D., Egger, M., Davidoff, F., Elbourne, D., Gøtzsche, P. C., and Lang, T., for the CONSORT Group. The revised consort statement for reporting randomized trials explanation and elaboration. *Annals Internal Med.* 2001; 134: 663–694.

Bailar, J. C. and Mosteller, F. Guidelines for statistical reporting in articles for medical journals. Amplifications and explanations. *Annals Internal Med.* 1988; 108: 66–73.

Begg, C. B., Cho, M., Eastwood, S., Horton, R., Moher, D., Olkin, I., Pitkin, R., Rennie, D., Schulz, K. F., Simel, D., and Stroup, D. F. Improving the quality of reporting of randomized controlled trials: the CONSORT Statement. *JAMA.* 1996; 276: 637–639.

Berger, V. W. Improving the information content of endpoints in clinical trials. *Controlled Clin Trials.* 2002; 23: 502–514.

Berger, V. W., Permutt, T., and Ivanova, A. Convex hull test of ordered categorical data. *Biometrics.* 1998; 54: 1541–1550.

Bickel, P., Klassen, C. A., Ritov, Y., and Wellner, J. *Efficient and Adaptive Estimation for Semiparametric Models.* Johns Hopkins University Press: Baltimore. 1993.

Bland, J. M. and Altman, D. G. Comparing methods of measurement: Why plotting difference against standard method is misleading. *Lancet.* 1995; 346: 1085–1087.

Bly, R. W. *Power-Packed Direct Mail: How to Get More Leads and Sales by Mail.* New York: Henry Holt. 1996.

Bly, R. W. *The Copywriter's Handbook: A Step-By-Step Guide to Writing Copy That Sells.* New York: Henry Holt. 1990.

Boyd, J. C., Rifari, N., and Annesley, T. M. Preparation of manuscripts for publication: improving your chances for success. *Clin Chem.* 2009; 55: 1259–1264.

Breslow, N. E. and Day, N. E. Statistical methods in cancer research, vol. 1: The analysis of case-control studies. *IARC Sci. Publ.* 1980; 325–338.

Canty, A. J., Davison, A. C., Hinkley, D. V., and Ventura, V. Bootstrap diagnostics and remedies. *Canadian J Stat.* 2006; 34: 5–27.

Carroll, R. J. and Ruppert, D. *Transformation and Weighting in Regression*. London: Chapman & Hall. 2000.

Cleveland, W. S. The *Elements of Graphing Data*. Hobart Press: Summit, NJ. 1994.

Converse, J. M. and Presser, S. *Survey Questions: Handcrafting the Standardized Questionnaire*. Thousand Oaks, CA: Sage. 1986.

Cox, D. R. The role of significance tests. *Scand J Stat.* 1977; 4: 49–70.

Cummings, P. and Koepsell, T. D. Statistical and design issues in studies of groups. *Inj Prev* 2002; 8: 6–7.

de Jong, M. G., Pieters, R., and Fox, J. P. Reducing social desirability bias through item randomized response: An application to measure under-reported desires. *J Marketing Res.* 2010; 47: 14–27.

Diciccio, T. J. and Romano, J. P. A review of bootstrap confidence intervals (with discussion). *JRSS B.* 1988; 50: 338–354.

Duggan, T. J. and Dean, C. W. Common misinterpretations of significance levels in sociological journals. *Amer Sociologist.* 1968; February: 45–46.

Efron, B. Bootstrap confidence intervals: good or bad? (with discussion). *Psychol Bull.* 1988; 104: 293–296.

Efron, B. Six questions raised by the bootstrap. In R. LePage and L. Billard, eds., *Exploring the Limits of the Bootstrap.* New York: Wiley; 1992. 99–126.

Efron, B. and Tibshirani, R. *An Introduction to the Bootstrap.* New York: Chapman & Hall. 1993.

Ehrenberg, A. S. C. Rudiments of numeracy. *JRSS Series A.* 1977; 140: 277–297.

Feng, Z., Diehr, P., Peterson, A., and McLerran, D. Selected statistical issues in group randomized trials. *Annu Rev Public Health.* 2001; 22: 167–187.

Finney, D. J. The *Theory of Experimental Design.* Chicago: University of Chicago Press. 1960.

Fisher, R. A. *Statistical Methods for Research Workers.* Ann Arbor, MI: University of Michigan Library. 1925; 1944.

Fisher, R. A. The *Design of Experiments.* London: Macmillan Pub Co. 1935; 1971.

Fowler, F. J., Jr. Designing questions to be good measures. In F. J. Fowler, *Survey Research Methods* (3rd ed.). Thousand Oaks, CA: Sage. 2002. pp. 76–103.

Freedman, D. A. A note on screening regression equations. *Amer Stat.* 1983; 37: 152–155.

Friedman, L. M., Furberg, C. D., and DeMets, D. L. *Fundamentals of Clinical Trials* (3rd ed.). St. Louis: Mosby. 1996.

Gail, M. H., Mark, S. D., Carroll, R., Green, S., and Pee, D. On design considerations and randomization-based inference for community intervention trials. *Stat. Med.* 1996; 15: 1069–92.

Good, P. I. *A Practitioner's Guide to Resampling Methods.* Boca Raton, FL: CRC Press. 2011.

Good, P. I. Robustness of Pearson Correlation. http://interstat. statjournals.net/YEAR/2009/articles/0906005.pdf.

Good, P. I. A new look at old inflationary theory. *Phys Essays.* 2010; 23: 368.

Good, P. I. *Managers' Guide to the Design and Conduct of Clinical Trials* (2nd ed.). New York: Wiley. 2006.

Good, P. I. and Hardin, J. *Common Errors in Statistics* (3rd ed.). New York: Wiley. 2009.

Good, P. I. and Lunneborg, L. Limitations of the analysis of variance. The one-way design. *J Modern Appl Stat Methods.* 2006; 5: 41–43.

Good, P. I. and Xie, F. Analysis of a crossover clinical trial by permutation methods. *Contemp Clin Trials* 2008; 29: 565–568.

Grant, A. Reporting controlled trials. *British J. Obstet Gynaecology.* 1989; 96: 397–400.

Hardin, J. W. and Hilbe, J. M. *Generalized Estimating Equations.* Chapman & Hall/CRC: London. 2003.

Hardin, J. W. and Hilbe, J. M. *Generalized Linear Models and Extensions* (3rd ed.). Stata Press: College Station, TX. 2012.

Hilbe, J. M. *Logistic Regression Models.* London: Chapman & Hall/CRC. 2009.

Hilbe, J. M. *Negative Binomial* Regression (2nd ed.). Cambridge University Press, 2011.

Hosmer, D. W. and Lemeshow, S. *Applied Logistic Regression* (2nd ed.). Wiley: New York. 2000.

Hosmer, D. W., Lemeshow, S., and May, S. *Applied Survival Analysis: Regression Modeling of Time to Event Data* (2nd ed.). New York: Wiley. 2008.

Husted, J. A., Cook, R. J., Farewell, V. T., and Gladman, D. D. Methods for assessing responsiveness: A critical review and recommendations. *J Clin Epidemiol.* 2000; 53: 459–468.

Huber, P. J. *Robust Statistics.* Wiley: New York. 1981.

Jagers, P. Invariance in the linear model. An argument for χ^2 and F in nonnormal situations. *Math. Operatonsforsch. Stat.* 1980; 11: 455–464.

John, L. K., Loewenstein, G., and Prelec, D. Measuring the prevalence of questionable research practices with incentives for truth-telling. *Psychol. Sci.* (forthcoming).

Kaplan, J. Misuses of statistics in the study of intelligence: the case of Arthur Jensen (with disc). *Chance.* 2001; 14: 14–26.

Lang, T. A. and Secic, M. *How to Report Statistics in Medicine.* American College of Physicians. Philadelphia.1997.

Lehmann, E. L. *Testing Statistical Hypotheses* (2nd ed.). New York: John Wiley & Sons. 1986.

Lehmann, E. L. and Casella, G. *Theory of Point Estimation* (2nd ed.). New York: Springer. 1998.

Leung, W. How to design a questionnaire. http://student.bmj.com/student/view-article.html?id=sbmj0106187.

Machin, D., Campbell, M. J., Tan, S.-B., and Tan, S.-H. *Sample Size Tables for Clinical Studies* (3rd ed.). London: BMJ Books. 2008

Mangel, M. and Samaniego, F. J. Abraham Wald's work on aircraft survivability. *JASA.* 1984; 79: 259–267.

Maritz, J. S. *Distribution Free Statistical Methods* (2nd ed.). London: Chapman & Hall. 1996.

Marshall, B. D. L., Milloy, M.-J., Wood, E., Montaner, J. S. G., and Kerr, T. Reduction in overdose mortality after the opening of North America's first medically supervised safer injecting facility: A retrospective population-based study. *The Lancet.* 2011; 377: 1429- 1437.

Martin, R. F. General Deming regression for estimating systematic bias and its confidence interval in method-comparison studies. *Clin Chem.* 2000; 46: 100–104.

Mathews, P. *Sample Size Calculations: Practical Methods for Engineers and Scientists.* State College, PA: Mathews Malnar and Bailey, Inc. 2010.

Mayo, D. G. *Error and the Growth of Experimental Knowledge.* Chicago: University of Chicago Press. 1996.

Mielke, P. W., Berry, K. J., Landsea, C. W., and Gray, W. M. Artificial skill and validation in meteorological forecasting. *Weather Forecast.* 1996; 11: 153–169.

Mosteller, F. and Tukey, J. W. *Data Analysis and Regression: A Second Course in Statistics.* Menlo Park, CA: Addison–Wesley. 1997.

Nardi, P. Developing a questionnaire. In *Doing Survey Research: A Guide to Quantitative Methods* (2nd ed., Chapter 4). Boston, MA: Pearson. 2006.

Neyman, J. Frequentist probability and frequentist statistics. *Synthese.* 1977; 36: 97–131.

Neyman, J. and Pearson, E. S. On the testing of specific hypotheses in relation to probability a priori. *Proc. Cambridge Phil. Soc.* 1933; 29: 492–510.

Neyman, J. and Scott, E. L. A theory of the spatial distribution of galaxies. *Astrophysical J.* 1952; 116: 144.

Ostapczuk, M., Moshagen, M., Zhao, Z., and Musch, J. Assessing sensitive attributes using the randomized-response-technique: Evidence for the importance of response symmetry. *J Educat Behav Stat.* 2009; 34: 267–287.

Parfrey, P. and Barrett, B. *Clinical Epidemiology: Practice and Methods.* New York: Humana Press. 2009.

Raab, G. M., Day, S., and Sales, J. How to select covariates to include in the analysis of a clinical trial. *Control Clin Trials.* 2000; 21: 330–342.

Schenker, N. Qualms about bootstrap confidence intervals. *JASA.* 1985; 80: 360–361.

Schroeder, Y. C. The procedural and ethical ramifications of pretesting survey questions. *Amer J Trial Advocacy.* 1987; 11: 195–201.

Selike, T., Bayarri, M. J., and Berger, J. O. Calibration of p-values for testing precise null hypotheses. *Amer Stat.* 2001; 55: 62–71.

Shao, J. and Tu, D. *The Jackknife and the Bootstrap.* New York: Springer. 1995.

Smith, P. G. and Douglas, A. J. Mortality of workers at the Sellafield plant of British Nuclear Fuels. *BMJ.* 1986; 293: 845–854.

Spector, P. *Data Manipulation with R.* New York: Springer. 2008.

Stöckl, D., Dewitte, K., and Thienpont, L. M. Validity of linear regression in method comparison studies: Is it limited by the statistical model or the quality of the analytical input data? *Clin Chem.* 1998; 44: 2340–2346.

Stone, A., Shiffman, S., Atienza, A., and Neibling L. *The Science of Real-Time Data Capture: Self-Reports in Health Research.* New York: Oxford University Press. 2007.

Strasak, A. M., Zaman, Q., Pfeiffer, K. P., Göbel, G., and Ulmer, H. Statistical errors in medical research—a review of common pitfalls. *Swiss Med Wkly.* 2007; 137: 44–49.

Therneau, T. M. and Grambsch, P. M. *Modeling Survival Data.* New York: Springer-Verlag. 2000.

Thomas, D. C. *Statistical Methods in Environmental Epidemiology.* New York: Oxford University Press. 2009.

Tufte, E. R. *The Visual Display of Quantitative Information.* Cheshire, CT: Graphics Press. 1983.

van Belle, G. *Statistical Rules of Thumb.* New York: Wiley. 2002.

Wald, A. *Statistical Decision Functions.* New York : Wiley. 1950.

Wilkinson, L. *The Grammar of Graphics.* New York: Springer-Verlag. 1999.

Author Index

Subject Index

R Function Index

Function Name	Purpose	Chapter
abs	take absolute value of	8
anova	format output of `lm()` as anova table	8
aov	fit an Analysis of Variance model	8
apply	functions over array margins	9
as.numeric	treat as number	9
attach	permit variables in dataframe to be accessed	8
barplot	create a barplot	17
boot::boot	bootstrap	7
boot::boot.ci	confidence interval	7
boxplot	boxplot	7
c or cat	concatenate in a vector	7
coef	regression coefficients	13
con, load	download data from a URL	8
cor.test	Pearson correlation	8
confit	compute confidence interval for coefficient	12
cov	covariance	13
survival:coxph	compute Cox hazard function	13
crossdes: MOLS	Latin Square designs	4
crossdes:BIB	incomplete blocks	4
crossdes:gen.factorial	factorial designs	4
data.frame	compile into a dataframe	8
dim	recast vector as a matrix	8
dim	dimensions of matrix	9
edit	modify a dataframe	8
equivalence::tost	test for equivalence	10
factor	treat as character	12
fisher.test	Fisher's exact test	8
fitted	fit a straight line to data	12
for	repeat calculations until	7
glm	general linear model	12
if, if … else	conditional command	8
I	forces glm to treat +,* as arithmetic operators	12
ifelse	conditional command	8
legend	create a legend on a graph	17

Function Name	Purpose	Chapter
length	number of items in a vector	7
library	load package in memory	8
lines	add line to existing scatter plot	12
lm	least squares regression	8
log	take logarithm	8
Rweka::logistic	logistic classification	13
psy:mdpsa	principal component analysis	3
mean	compute mean of a vector	8
median	compute median	7
mtext	add title to graph	17
naclus	detect clusters of missing data	6
ncol	number of columns in matrix	9
nrow	number of rows in matrix	9
numeric	reserve space	7
pbinom	binomial cumulative probability	8
pnbinom	negative binomial cumulative probability	8
plot	scatter plot	12
power.anova.test	computes power of analysis of variance	11
power.t.test	computes power of t-test	11
ppoints	generate sequence of probability points	17
predict	compute confidence bounds for line	12
qbinom	binomial quantile	8
qnbinom	negative binomial quantile	8
qpois	Poisson quantile	8
quantile	compute quantiles of sample	10
read.table	input a table from an external source	8
gdata::remove.vars	remove variables from a data frame	3
leaps::regsubsets	subset stepwise regression	12
rep	repeat a value	8
require	give precedence to package	9
residuals	display residuals of glm	12
return	return from functions with this value	10
rnorm	generate values from a Normal distribution	17
rpart::rpart	build a decision tree	14
quantreg::rq	LAD, quantile regression	13
sample	sample or rearrange vector	7
solve	solve matrix equation	9
sort	sort vector	10
sqrt	square root	13

Function Name	**Purpose**	**Chapter**
step	stepwise regression	14
stripchart	plot stripchart	7
MASS::stepAIC	stepwise regression	12
sum	sum vector components	8
summary	display results of function	12
survival::survfit	fit survival curve	13
symbols	plot 3rd variable as a symbol	17
t.test	student's t-test	8
equivalence::tost	test for equivalence	10
var	variance	7
vcd::goodfit	goodness of fit of a $1 \times N$ distribution	8
while	repeat calculations while	9
with	use given data to perform a function	17

For Product Safety Concerns and Information please contact our EU
representative GPSR@taylorandfrancis.com
Taylor & Francis Verlag GmbH, Kaufingerstraße 24, 80331 München, Germany

www.ingramcontent.com/pod-product-compliance
Lightning Source LLC
Chambersburg PA
CBHW060351220326
41598CB00023B/2876